INTRODUCTORY CHEMICAL ENGINEERING

Edited by
Rose Marie O. Mendoza

Delve Publishing

Introductory Chemical Engineering

Edited by Rose Marie O. Mendoza

Edition 2018

Delve Publishing

Published by

Delve Publishing LLC

708 3rd Avenue, 6th Floor

New York

NY 10801

United States of America

www.delvepublishing.com

ISBN: 978-1-68095-908-6 (Hardback)

Printed in United States of America

Library of Congress Control Number: 2016920674

About the Editor

Rose Marie O. Mendoza

Dr. Rose obtained her PhD in Chemical Engineering from the University of the Philippines-Diliman in 2013. She is also a Professor in the Graduate School Department under the Master of Engineering Program at Adamson University since 2006, and is a Visiting Research Fellow at the Department of Environmental Engineering and Science, and the Department of Environmental Resource Management in Chia Nan University of Pharmacy and Science, Tainan Taiwan since 2010. She is now taking up her Post Doctorate Degree in Green Power: Hydrogen Generation and Fuel Cell Development at the University of Philippines-Diliman under the Department of Mining, Metallurgy and Materials Engineering and the University of California, Merced, USA under the Department of Mechanical Engineering.

Table of Contents

3 UNIT OPERATIONS

4 UNIT PROCESSES

5 CHEMICAL PROCESS DESIGN

6 CHEMICAL PROCESS

7 TRANSPORT PHENOMENA

8 CHEMICAL PLANT OPERATIONS

9 CHEMICAL THERMODYNAMICS

List of Abbreviations

ADP	Adenosine Diphosphate
AMP	Adenosine Monophosphate
ATP	Adenosine Triphosphate
BFD	Block Flow Diagrams
BLS	Bureau of Labor Statistics
CAST	Computing and Systems Technology
CRO	Contract Research Organizations
HFAU	Hydrofluoric Alkylation Unit
HMW	High Molecular Weight
ICCM	International Conference on Chemicals Management
KPI	Key Performance Indicators
LCA	Life Cycle Assessment
LMW	Low Molecular Weight
LPG	Liquefied Petroleum Gas
MRT	Microreactor Technology
NEPIC	Northeast of England Process Industry Cluster
ODH	Oxidative Dehydrogenation
PFD	Process Flow Diagram
PFR	Plug Flow Reactor
PHA	Process Hazard Analysis
PP	Polypropylene
PSE	Process Systems Engineering
PVC	Polyvinyl Chloride

RCRA	Resource Conservation and Recovery Act
REACH	Registration Evaluation Authorization and Restriction of Chemicals
SAAU	Sulfuric Acid Alkylation Unit
SAICM	Strategic Approach to International Chemicals Management
SOCMA	Society of Chemical Manufacturers and Affiliates
TSCA	Toxic Substances Control Act
UNEP	United Nations Environmental Programme
UNIDO	United Nations Industrial Development Organization
VCM	Vinyl Chloride Monomer

Preface

Chemical engineering is the branch of engineering that deals with chemical production and the manufacture of products through chemical processes, together with applied mathematics and economics to produce, transform, transport, and properly use chemicals, materials and energy. This includes designing equipment, systems and processes for refining raw materials and for mixing, compounding and processing chemicals to make valuable products. Chemical engineers are in great demand because of the large number of industries that depend on the synthesis and processing of chemicals and materials. In addition to traditional careers in the chemical, energy and oil industries, chemical engineers enjoy increasing opportunities in biotechnology, pharmaceuticals, electronic device fabrication and environmental engineering. The unique training of the chemical engineer becomes essential in these areas when processes involve the chemical or physical transformation of matter. For example, chemical engineers working in the chemical industry investigate the creation of new polymeric materials with important electrical, optical or mechanical properties. This requires attention not only to the synthesis of the polymer, but also to the flow and forming processes necessary to create a final product. In biotechnology, chemical engineers help design production facilities that use microorganisms and enzymes to synthesize new drugs. Problems in environmental engineering that engage chemical engineers include the development of processes (catalytic converters, effluent treatment facilities) to minimize the release of or deactivate products that are harmful to the environment. To do these jobs, chemical engineering students must have a complete and quantitative understanding of both the engineering and scientific principles underlying these technological processes.

Content Coverage

Chapters One and *Two* present coverage of chemical engineering and overview of chemical industry, respectively. This section also discusses the historical development of chemical engineering as a science and as a profession as well as the basic processes concerning the practice of chemical engineering.

Chapters Three and *Four* focuses on unit operations and processes, separately. In chemical engineering, unit operations involve a physical change or chemical transformation such as separation, crystallization, evaporation, filtration, polymerization, isomerization, and other reactions, but unit processes involves a chemical changes along with physical change.

Chapters Five and *Six* covers chemical engineering chemical process and process design, correspondingly. Process design is central to chemical engineering, and it can be considered to be the summit of that field, bringing together all of the field's components. Process design is distinct from equipment design, and is considered closer to operations. Processes often include many unit operations.

Chapters Seven and *Eight* clarify the fundamentals of transport phenomenon and chemical plant operations, separately. The study of transport phenomena concerns the exchange of mass, energy, and momentum between observed and studied systems, however chemical plant is designed for the purpose of processing and producing chemicals.

Chapter Nine is intended to focus on chemical thermodynamics that deals with the study of the interrelation of heat and work with chemical reactions or with physical changes of state within the confines of the laws of thermodynamics. Chemical thermodynamics involves not only laboratory measurements of various thermodynamic properties, but also the application of mathematical methods to the study of chemical questions and the spontaneity of processes.

CHAPTER 1

Chemical Engineering: An Overview

Chapter Contents

- **Introduction**
- **Etymology**
- **History**
- **Origin**
- **Definitions**
- **Concepts**
- **Applications and Practice**

Chemical engineering is the study and practice of transforming substances for the tangible improvement of the human condition. The heart of the chemical engineering lies mainly from the transformation of raw materials into useful products and its production for commercial and large-scale consumption.

Chemical Engineering differs from other related courses such as chemistry and the allied courses mainly in its focus on process design and optimization and its application to large-scale manufacturing. In addition, an industrial plant cannot be set-up without the participation of the chemical engineer primarily from conception to construction of the plant. The detailed process plant lay-out needs to be evaluated and approved by a chemical engineer before construction.

INTRODUCTION

Chemical engineering is a branch of engineering that applies physical sciences (physics and chemistry), life sciences (microbiology and biochemistry), together with applied mathematics and economics to produce, transform, transport, and properly use chemicals, materials and energy. A chemical engineer designs large-scale processes that convert chemicals, raw materials, living cells, microorganisms and energy into useful forms and products.

A distillation column. One of the many processes a Chemical Engineer designs, operate and control.

ETYMOLOGY

A 1996 *British Journal for the History of Science* article cites James F. Donnelly for mentioning an 1839 reference to chemical engineering in relation to the production of sulfuric acid. In the same paper however, George E. Davis, an English consultant, was credited for having coined the term.

The *History of Science in United States: An Encyclopedia* puts this at around 1890. "Chemical engineering", describing the use of mechanical equipment in the chemical industry, became common vocabulary in England after 1850. By 1910, the profession, "chemical engineer," was already in common use in Britain and the United States.

George E. Davis.

HISTORY

Chemical engineering as a discipline that was developed out of those practicing "industrial chemistry" in the late 19th century. Before the Industrial Revolution (18th century), industrial chemicals and other consumer products such as soap were mainly produced through batch processing. Batch processing is labor-intensive and individuals mix predetermined amounts of ingredients in a vessel, heat, cool or pressurize the mixture for a predetermined length of time. The product may then be isolated, purified and tested to achieve a saleable product. Batch processes are still performed today on higher value products, such as pharmaceutical intermediates, specialty and formulated products such as perfumes and paints, or in food manufacture such as pure maple syrups, where a profit can still be made despite batch methods being slower and inefficient in terms of labor and equipment usage. Due to the application of Chemical Engineering techniques during manufacturing process development, larger volume chemicals are now produced through a continuous "assembly line" chemical processes. The Industrial Revolution was when a shift from batch to more continuous processing began to occur. Today commodity chemicals and petrochemicals are predominantly made using continuous manufacturing processes whereas specialty chemicals, fine chemicals and pharmaceuticals are made using batch processes.

ORIGIN

The Industrial Revolution led to an unprecedented escalation in demand, both with regard to quantity and quality, for bulk chemicals such as soda ash. This meant two things: one, the size of the activity and the efficiency of operation had to be enlarged, and two, serious alternatives to batch processing, such as continuous operation, had to be examined.

The First Chemical Engineer

Industrial chemistry was being practiced in the mid 1800s, but it was not until the 1880s that the engineering elements required to control chemical processes were being recognized as a distinct professional activity. Chemical engineering was first established as a profession in the United Kingdom when the first chemical engineering course was given at the University of Manchester in 1887 by George E. Davis in the form of twelve lectures covering various aspects of industrial chemical practice. As a consequence George E. Davis is regarded as the world's first chemical engineer. Today, chemical engineering is a highly regarded profession. Chemical engineers with experience can become licensed Professional Engineers in the United States, aided by the National Society of Professional Engineers, or gain "Chartered" chemical-engineer status through the UK-based Institution of Chemical Engineers.

Professional Associations

In 1880, the first attempt was made to form a *Society of Chemical Engineers* in London. This eventually resulted in the formation of the Society of Chemical Industry in 1881. The American Institute of Chemical Engineers (AIChE) was founded in 1908, and the UK Institution of Chemical Engineers (IChemE) in 1922. These substantial international membership. Some other countries now have chemical engineering societies or sections within chemical or engineering societies, but the AIChE and IChemE remain the major ones in numbers and international spread: they are both open to suitably qualified professionals or students of chemical engineering anywhere in the world.

DEFINITIONS

For the other established branches of engineering, there were ready associations in the mind of the common man: Mechanical Engineering meant machines, Electrical Engineering meant circuitry, and Civil Engineering meant structures. So chemical engineering can be associated to chemicals.

Unit Operation

The answer, provided by Arthur D. Little to the president of MIT, was to emphasize the approach chemical engineers took to the design and analysis of processes rather than a process or a product. The concept of Unit operations was developed to emphasize the underlying unity among seemingly different operations. For example, the principles are the same whether one is concerned about separating alcohol from water in a fermenter, or separating gasoline from diesel in a refinery, as long as the basis of separation is generation of a vapor of different composition from the liquid. Therefore, such separation processes can be studied together as a unit operation (in this case called distillation). The concept has given the profession a good stand in its phase of growth, and has even been used to understand the way the human body functions.

Unit Processes

In the early part of the last century, a parallel concept called Unit Processes was used to classify reactive processes. Thus oxidations, reductions, alkylations, etc. formed separate unit processes and were studied independently. This was natural considering the close affinity of chemical engineering to industrial chemistry at its inception. Gradually however, the subject of chemical reaction engineering has largely replaced the unit process concept. This subject looks at the entire body of chemical reactions as having a personality of its own, independent of the particular chemical species or chemical bonds involved. In the design and process control of chemical reactors, a knowledge on the rate behavior, thermodynamics, single or multi-phase reactions as well as thorough understanding of chemical bonding and reactions are required for the efficient and effective practice of chemical engineering as a profession. The emergence of chemical reaction engineering as a discipline truly signaled the severance of the umbilical cord connecting chemical engineering to industrial chemistry, and served to cement the truly unique character of this discipline.

CONCEPTS

Chemical engineering involves the application of several principles. Key concepts are presented below.

Chemical Reaction Engineering

Chemical engineering involves managing plant processes and conditions to ensure optimal plant operation. Chemical reaction engineers construct models for reactor analysis and design using laboratory data and physical parameters, such as chemical thermodynamics, to solve problems and predict reactor performance.

Plant Design and Construction

Chemical engineering design is concerned on the construction of plans, specification, and economic analyses for pilot plants, new plants or plant modifications. Design engineers often work in a consulting role, designing plants to meet clients' needs. Design is limited by a number of factors, including funding, government regulations and safety standards. These constraints dictate a plant's choice of process, materials and equipment.

Plant construction is coordinated by project engineers and project managers depending on the size of the investment. A chemical engineer may do the job of project engineer full-time or part of the time, which requires additional training and skills, or act as a consultant to the project group. In USA the education of chemical engineering graduates from the Baccalaureate programs accredited by Accreditation Board for Engineering and Technology (ABET) do not usually stress project engineering education, which can be obtained by specialized training, as electives, or from graduate programs. Project engineering jobs are some of the largest employers for chemical engineers.

Process Design and Analysis

A unit operation is a physical step in an individual chemical engineering process. Unit operations (such as crystallization, filtration, drying and evaporation) are used to prepare reactants, purifying and separating its products, recycling unspent reactants, and controlling energy transfer in reactors. On the other hand, a unit process is the chemical equivalent of a unit operation. Along with unit operations, unit processes constitute an operation. Unit processes (such as nitration and oxidation) involve the conversion of material by biochemical, thermochemical and other means. Chemical engineers responsible for these are called process engineers.

Process design requires the definition of equipment types and sizes as well as how they are connected together and the materials of construction. Details are often printed on a Process Flow Diagram which is used to control the capacity and reliability of a new or modified chemical factory.

Education for chemical engineers in the first college degree 3 or 4 years of study stresses the principles and practices of process design. The same skills are used in existing chemical plants to evaluate the efficiency and make recommendations for improvements.

Transport Phenomena

Modeling and analysis of transport phenomena is essential for many industrial applications. Transport phenomena involve fluid dynamics, heat transfer and mass transfer, which are governed mainly by momentum transfer, energy transfer and transport of chemical species respectively. Models often involve separate considerations for macroscopic, microscopic and molecular level phenomena. Modeling of transport phenomena requires therefore an understanding of applied mathematics.

APPLICATIONS AND PRACTICE

Operators in a chemical plant using an older analog control board, seen in East-Germany, 1986.

Chemical engineers "develop economic ways of using materials and energy". Chemical engineers use chemistry and engineering to turn raw materials into usable products, such as medicine, petrochemicals and plastics on a large-scale, industrial setting. They are also involved in waste management and research. In modern day practice, chemical engineering research, both basic and applied research are aided computers.

Chemical engineers use computers to control automated systems in plants.

Chemical engineers may be involved in industry or university research where they are tasked with designing and performing experiments to create better and safer methods for production, pollution control, and resource conservation. They may be involved in designing and constructing plants as a project engineer. Chemical Engineers, as Project Engineers use their knowledge and skills in selecting optimum production methods and cost-effective equipments that will maximize safety as well as company profitability. After plant construction, chemical engineering project managers may be involved in equipment upgrades, process changes, troubleshooting, and daily operations in either full-time or consulting roles.

REFERENCES

1. Cohen 1996, p. 172.
2. Cohen 1996, p. 174.
3. Reynolds 2001, p. 176.
4. Cohen 1996, p. 186.
5. Perkins 2003, p. 20.
6. Cohen 1996, pp. 172–173.

7. Cohen 1996, p. 175.
8. Cohen 1996, p. 178.
9. Cohen 1996, p. 180.
10. Cohen 1996, p. 183.
11. Cohen 1996, p. 184.
12. Cohen 1996, p. 187.
13. Cohen 1996, p. 189.
14. Cohen 1996, p. 190.
15. Cohen 1996, p. 185.
16. Ogawa 2007, p. 2.
17. Perkins 2003, p. 29.
18. Perkins 2003, p. 30.
19. Perkins 2003, p. 31.
20. Reynolds 2001, p. 177.
21. Perkins 2003, pp. 32–33.
22. Kim 2002, p. 7S.
23. Kim 2002, p. 8S.
24. Perkins 2003, p. 35.
25. Kim 2002, p. 9S.
26. American Institute of Chemical Engineers 2003a.
27. Towler & Sinnott 2008, pp. 2–3.
28. Herbst, Andrew; Hans Verwijs (Oct. 19-22). "Project Engineering: Interdisciplinary Coordination and Overall Engineering Quality Control". Proc. of the Annual IAC conference of the American Society for Engineering Management 1 (ISBN 9781618393616): 15–21
29. "What Do Chemical Engineers Do?".
30. McCabe, Smith & Hariott 1993, p. 4.
31. Silla 2003, pp. 8–9.
32. Bird, Stewart & Lightfoot 2002, pp. 1–2.
33. Garner 2003, pp. 47–48.
34. American Institute of Chemical Engineers 2003, Article III.
35. Garner 2003, pp. 49–50.
36. Delgass; et al. "Seventy-Five Years of Chemical Engineering". Purdue University. Retrieved 13 August 2013.
37. W. F. Furter (1980) A Century of Chemical Engineering, Plenum Press (NY & London) ISBN 0-306-40895-3

CHAPTER 2

Chemical Industry

Chapter Contents

- **Introduction**
- **History**
- **Products**
- **Technology**
- **Chemical Leasing**

The chemical industry is an industry where its raw materials are distinguished more for their chemical properties than for their physical or mechanical properties. It processes these raw material into higher added value chemical products. The processing may be by increased purity, increased concentration, chemical reactions or a combination of the above.

The chemical industry is characterized by the high sophistication of its processes, by advanced automatic controls of plant operations and of products quality, by a high investment per employee, and by high productivity per employee. These main attributes of a chemical industry is also the main reasons why they need individuals with good qualifications and professional ability.

INTRODUCTION

The chemical industry comprises the companies that produce industrial chemicals. Central to the modern world economy, it converts raw materials (oil, natural gas, air, water, metals, and minerals) into more than 70,000 different products.

In the plastic industry, for instance, overlaps are unavoidable, since the companies does not solely produce plastics-they also other chemicals as side and/or by-products.

Oil refinery in Louisiana - an example of chemical industry.

HISTORY

Although chemicals were made and used throughout history, the birth of the heavy chemical industry (production of chemicals in large quantities for a variety of uses) coincided with the beginnings of the Industrial Revolution in general.

Industrial Revolution

One of the first chemicals to be produced in large amounts through industrial process was sulfuric acid. In 1736, the pharmacist Joshua Ward developed a process for its production that involved heating saltpeter, allowing the sulfur to oxidize and combine with water. It was the first practical production of sulfuric acid on a large scale. John Roebuck and Samuel Garbett were the first to establish a large-scale factory in Prestonpans, Scotland, in 1749, which used leaden condensing chambers for the manufacture of sulfuric acid.

Charles Tennant's St. Rollox Chemical Works in 1831, then the biggest chemical enterprise in the world.

In the early 18th century, cloth was bleached by treating it with stale urine or sour milk and exposing it to sunlight for long periods of time, which created a severe bottleneck in production. Sulfuric acid began to be used as a more efficient agent as well as lime by the middle of the century, but it was the discovery of bleaching powder by Charles Tennant that spurred the creation of the first great chemical industrial enterprise. His powder was made by reacting chlorine with dry slaked lime and proved to be a cheap and successful product. He opened a factory in St Rollox, north of Glasgow, and production went from just 52 tons in 1799 to almost 10,000 tons just five years later.

Soda ash was used since ancient times in the production of glass, textile, soap, and paper, and the source of the potash had traditionally been wood ashes in Western Europe. By the 18th century, this source was becoming uneconomical due to deforestation, and the French Academy of Sciences offered a prize of 2400 livres for a method to produce alkali from sea salt (sodium chloride). In 1791, Nicolas Leblanc introduced the Leblanc process of making soda ash (sodium caronate) which involved two stages: production of sodium sulfate from sodium chloride, followed by reaction of the sodium sulfate with coal and calcium carbonate to produce sodium carbonate. However, no price was given to Leblanc during that time due to the spark of the French Revolution.

However, it was in Britain that the Leblanc process really took off. William Losh built the first soda works in Britain at the Losh, Wilson and Bell works on the River Tyne in 1816, but it remained on a small scale due to large tariffs on salt production until 1824. When these tariffs were repealed, the British soda industry was able to rapidly expand. James Muspratt's chemical works in Liverpool and Charles Tennant's complex near Glasgow became the largest chemical production centres anywhere. By the 1870s, the British soda output of 200,000 tons annually exceeded that of all other nations in the world combined.

Ernest Solvay, patented an improved industrial method for the manufacture of soda ash

These huge factories began to produce a greater diversity of chemicals as the Industrial Revolution matured. Originally, large quantities of alkaline waste were vented into the environment from the production of soda, provoking one of the first pieces of environmental legislation to be passed in 1863. This provided for close inspection of the factories and imposed heavy fines on those exceeding the limits on pollution. Methods were soon devised to make useful byproducts from the alkali.

The Solvay process was developed by the Belgian industrial chemist Ernest Solvay in 1861. In 1864, Solvay and his brother Alfred constructed a plant in the Belgian town of Charleroi and in 1874, they expanded into a larger plant in Nancy, France. The new process proved more economical and less polluting than the Leblanc method, and its use spread. In the same year, Ludwig Mond visited Solvay to acquire the rights to use his process, and he and John Brunner formed the firm of Brunner, Mond & Co., and built a Solvay plant at Winnington, England. Mond was instrumental in making the Solvay process a commercial success; he made several refinements between 1873 and 1880 that removed byproducts that could slow or halt the mass production of sodium carbonate through use of the process.

Expansion and Maturation

The late 19th century saw an explosion in both the quantity of production and the variety of chemicals that were manufactured. Large chemical industries also took shape in Germany and later in the United States.

The factories of the German firm BASF, in 1866.

Production of artificial manufactured fertilizer for agriculture was pioneered by Sir John Lawes at his purpose-built Rothamsted Research facility. In the 1840s he established large works near London for the manufacture of superphosphate of lime. Processes for the vulcanization of rubber were patented by Charles Goodyear in the United States and Thomas Hancock in England in the 1840s. The first synthetic dye was discovered by William Henry Perkin in London. He partly transformed aniline into a crude mixture which, when extracted with alcohol, produced a substance with an intense purple color.

He also developed the first synthetic perfumes. However, it was the German industry that quickly began to dominate the field of synthetic dyes. The three major firms BASF, Bayer and Hoechst produced several hundred different dyes, and by 1913, the German industry produced almost 90 percent of the world supply of dyestuffs and sold about 80 percent of their production abroad. In the United States, Herbert Henry Dow's use of electrochemistry to produce chemicals from brine was a commercial success that helped to promote the country's chemical industry.

The petrochemical industry can be traced back to the oil works of James Young in Scotland and Abraham Pineo Gesner in Canada. The first plastic was invented by Alexander Parkes, an English metallurgist. In 1856, he patented Parkesine, a celluloid based on nitrocellulose treated with a variety of solvents. This material, exhibited at the 1862 London International Exhibition, anticipated many of the modern aesthetic and utility uses of plastics. The industrial production of soap from vegetable oils was started by William Lever and his brother James in 1885 in Lancashire based on a modern chemical process invented by William Hough Watson that used glycerin and vegetable oils.

By the 1920s, chemical firms consolidated into large conglomerates; IG Farben in Germany, Rhône-Poulenc in France and Imperial Chemical Industries in Britain. Dupont became a major chemicals firm in the early 20th century in America.

Currently chemical production is a high-tech industry, where the competitiveness is more based on capacity in investment on research and development than the labor cost.

PRODUCTS

Polymers and plastics, especially polyethylene, polypropylene, polyvinyl chloride, polyethylene terephthalate, polystyrene and polycarbonate comprise about 80% of the industry's output worldwide. These materials are often converted to fluoropolymer tubing products and used by the industry to transport highly corrosive materials. Chemicals are used to make a wide variety of consumer goods, as well as thousands of inputs to agriculture, manufacturing, construction, and service industries. The chemical industry itself consumes 26 percent of its own output. Major industrial customers include rubber and plastic products, textiles, apparel, petroleum refining, pulp and paper, and primary metals. Chemicals are nearly a $3 trillion global enterprise, and the EU and U.S. chemical companies are the world's largest producers.

Sales of the chemical business can be divided into a few broad categories, including basic chemicals (about 35 to 37 percent of the dollar output), life sciences (30 percent), specialty chemicals (20 to 25 percent) and consumer products (about 10 percent).

Basic Chemicals & Commodity Chemicals to Polymers and Specialty Chemicals

Basic chemicals, or "commodity chemicals" are a broad chemical category including polymers, bulk petrochemicals and intermediates, other derivatives and basic industrials, inorganic chemicals, and fertilizers. Typical growth rates for basic chemicals are about 0.5 to 0.7 times GDP. Product prices are generally less than fifty cents per pound.

New polypropylene plant PP3 in the Slovnaft oil refinery (Bratislava, Slovakia).

Polymers, the largest revenue segment at about 33 percent of the basic chemicals dollar value, includes all categories of plastics and man-made fibers. The major markets for plastics are packaging, followed by home construction, containers, appliances, pipe, transportation, toys, and games.

- The largest-volume polymer product, polyethylene (PE), is used mainly in packaging films and other markets such as milk bottles, containers, and pipe.
- Polyvinyl chloride (PVC), another large-volume product, is principally used to make pipe for construction markets as well as siding and, to a much smaller extent, transportation and packaging materials.
- Polypropylene (PP), similar in volume to PVC, is used in markets ranging from packaging, appliances, and containers to clothing and carpeting.
- Polystyrene (PS), another large-volume plastic, is used principally for appliances and packaging as well as toys and recreation.
- The leading man-made fibers include polyester, nylon, polypropylene, and acrylics, with applications including apparel, home furnishings, and other industrial and consumer use.

The principal raw materials for polymers are bulk petrochemicals.

Chemicals in the bulk petrochemicals and intermediates are primarily made from liquefied petroleum gas (LPG), natural gas, and crude oil. Their sales volume is close to 30 percent of overall basic chemicals. Typical large-volume products include ethylene, propylene, benzene, toluene, xylenes, methanol, vinyl chloride monomer (VCM), styrene, butadiene, and ethylene oxide. These basic or commodity chemicals are the starting materials used to manufacture many polymers and other more complex organic chemicals particularly those that are made for use in the specialty chemicals category.

Other derivatives and basic industrials include synthetic rubber, surfactants, dyes and pigments, turpentine, resins, carbon black, explosives, and rubber products and contribute about 20 percent of the basic chemicals' external sales.

Inorganic chemicals (about 12 percent of the revenue output) make up the oldest of the chemical categories. Products include salt, chlorine, caustic soda, soda ash, acids (such as nitric acid, phosphoric acid, and sulfuric acid), titanium dioxide, and hydrogen peroxide.

Fertilizers are the smallest category (about 6 percent) and include phosphates, ammonia, and potash chemicals.

Life Sciences

Life sciences (about 30 percent of the dollar output of the chemistry business) include differentiated chemical and biological substances, pharmaceuticals, diagnostics, animal health products, vitamins, and pesticides. While much smaller in volume than other chemical sectors, their products tend to have very high prices—over ten dollars per pound—growth rates of 1.5 to 6 times GDP, and research and development spending at 15 to 25 percent of sales. Life science products are usually produced with very high specifications and are closely scrutinized by government agencies such as the Food and Drug Administration. Pesticides, also called "crop protection chemicals", are about 10 percent of this category and include herbicides, insecticides, and fungicides.

Specialty Chemicals

Specialty chemicals are a category of relatively high valued, rapidly growing chemicals with diverse end product markets. Typical growth rates are one to three times GDP with prices over a dollar per pound. They are generally characterized by their innovative aspects. Products are sold for what they can do rather than for what chemicals they contain. Products include electronic chemicals, industrial gases, adhesives and sealants as well as coatings, industrial and institutional cleaning chemicals, and catalysts. In 2012, excluding fine chemicals, the $546 billion global specialty chemical market was 33% Paints, Coating and Surface Treatments, 27% Advanced Polymer, 14% Adhesives and Sealants, 13% additives and 13% pigments and inks.

Specialty chemicals are sold as effect or performance chemicals. Sometimes they are mixtures of formulations, unlike "fine chemicals," which are almost always single-molecule products.

Consumer Products

Consumer products include direct product sale of chemicals such as soaps, detergents, and cosmetics. Typical growth rates are 0.8 to 1.0 times GDP.

Consumers rarely if ever come into contact with basic chemicals but polymers and specialty chemicals are the materials that they will encounter everywhere in their everyday lives, such as in plastics, cleaning materials, cosmetics, paints & coatings, electronic gadgets, automobiles and the materials used to construct their homes. These specialty products are marketed by chemical companies to the downstream manufacturing industries as pesticides, specialty polymers, electronic chemicals, surfactants, construction chemicals, Industrial Cleaners, flavors and fragrances, specialty coatings, printing inks, water-soluble polymers, food additives, paper chemicals, oil field chemicals, plastic adhesives, adhesives and sealants, cosmetic chemicals, water management chemicals, catalysts, textile chemicals. Chemical companies rarely supply these products directly to the consumer.

Every year, the American Chemistry Council tabulates the U.S. production volume of the top 100 basic chemicals. In 2000, the aggregate production volume of the top 100 chemicals totalled 502 million tons, up from 397 million tons in 1990. Inorganic chemicals tend to be the largest volume, though much smaller in dollar revenue terms due to their low prices. The top 11 of the 100 chemicals in 2000 were sulfuric acid (44 million tons), nitrogen (34), ethylene (28), oxygen (27), lime (22), ammonia (17), propylene (16), polyethylene (15), chlorine (13), phosphoric acid (13) and diammonium phosphates (12).

Companies

The largest chemical producers today are global companies with international operations and plants in numerous countries. A list of the top 25 chemical companies by chemical sales in 2015 appears below. (Note: Chemical sales represent only a portion of total sales for some companies.)

Top chemical companies by chemical sales in 2015.

Rank	Company	2015 Chemical Sales (USD in billions)	Headquarters
1	BASF	$63.7	Ludwigshafen, Germany
2	Dow Chemical Company	$48.8	Midland, Michigan, United States
3	Sinopec (China Petroleum & Chemical Corporation)	$43.8	Beijing, China
4	SABIC (Saudi Arabia Basic Industries Corporation)	$34.3	Riyadh, Saudi Arabia
5	Formosa Plastics Corporation	$29.2	Kaohsiung City, Taiwan
6	INEOS	$28.5	Rolle, Switzerland
7	ExxonMobil Corp.	$28.1	Irving, Texas, United States

8	LyondellBasell	$26.7	Houston, Texas, United States, and London, United Kingdom
9	Mitsubishi Chemical	$24.3	Tokyo, Japan
10	DuPont	$20.7	Wilmington, Delaware, United States
11	LG Chem	$18.2	Seoul, South Korea
12	Air Liquide	$17.3	Paris, France
13	The Linde Group	$16.8	Munich, Germany
14	Akzo Nobel	$16.5	Amsterdam, Netherlands
15	Toray Industries	$15.5	Tokyo, Japan
16	Evonik Industries	$15.0	Essen, Germany
17	PPG Industries	$14.2	Pittsburgh, Pennsylvania, United States
18	Braskem	$14.2	São Paulo, Brazil
19	Yara International	$13.9	Oslo, Norway
20	Covestro	$13.4	Leverkusen, Germany
21	Sumitomo Chemical	$13.3	Tokyo, Japan
22	Reliance Industries	$12.9	Mumbai, India
23	Solvay	$12.3	Brussels, Belgium
24	Bayer	$11.5	Leverkusen, Germany
25	Mitsui Chemicals	$11.1	Tokyo, Japan

TECHNOLOGY

From the perspective of chemical engineers, the chemical industry involves the use of chemical processes such as chemical reactions and refining methods to produce a wide variety of solid, liquid, and gaseous materials. Most of these products serve to manufacture other items, although a smaller number go directly to consumers. Solvents, pesticides, lye, washing soda, and portland cement provide a few examples of product used by consumers.

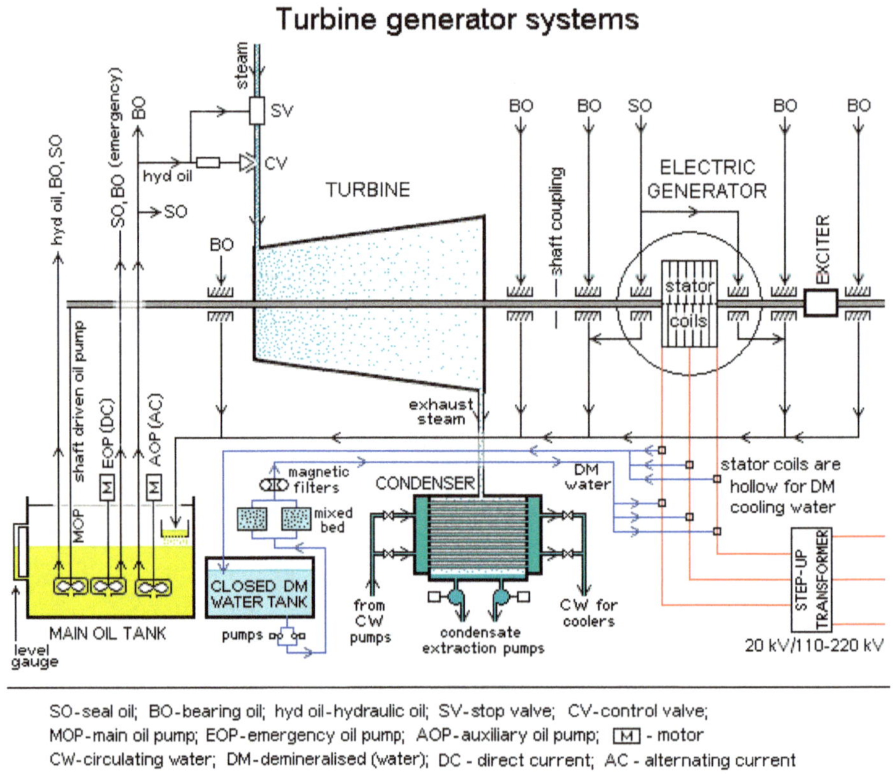

This is a process diagram of a turbine generator. Engineers working to produce a sustainable process for use in the chemical industry need to know how to design a sustainable process in which the system can withstand or manipulate process-halting conditions such as heat, friction, pressure, emissions, and contaminants.

The industry includes manufacturers of inorganic- and organic-industrial chemicals, ceramic products, petrochemicals, agrochemicals, polymers and rubber (elastomers), oleochemicals (oils, fats, and waxes), explosives, fragrances and flavors. Examples of these products are shown in the Table below.

Product Type	**Examples**
inorganic industrial	ammonia, chlorine, sodium hydroxide, sulfuric acid, nitric acid
organic industrial	acrylonitrile, phenol, ethylene oxide, urea
ceramic products	silica brick, frit
petrochemicals	ethylene, propylene, benzene, styrene
agrochemicals	fertilizers, insecticides, herbicides

polymers	polyethylene, Bakelite, polyester
elastomers	polyisoprene, neoprene, polyurethane
oleochemicals	lard, soybean oil, stearic acid
explosives	nitroglycerin, ammonium nitrate, nitrocellulose
fragrances and flavors	benzyl benzoate, coumarin, vanillin
industrial gases	nitrogen, oxygen, acetylene, nitrous oxide

Although the pharmaceutical industry is often considered a chemical industry , it has many different characteristics that puts it in a separate category. Other closely related industries include petroleum, glass, paint, ink, sealant, adhesive, and food processing manufacturers.

Chemical processes such as chemical reactions operate in chemical plants to form new substances in various types of reaction vessels. In many cases the reactions take place in special corrosion-resistant equipment at elevated temperatures and pressures with the use of catalysts. The products of these reactions are separated using a variety of techniques including distillation especially fractional distillation, precipitation, crystallization, adsorption, filtration, sublimation, and drying.

The processes and product or products are usually tested during and after manufacture by dedicated instruments and on-site quality control laboratories to ensure safe operation and to assure that the product will meet required specifications. More organizations within the industry are implementing chemical compliance software to maintain quality products and manufacturing standards. The products are packaged and delivered by many methods, including pipelines, tank-cars, and tank-trucks (for both solids and liquids), cylinders, drums, bottles, and boxes. Chemical companies often have a research-and-development laboratory for developing and testing products and processes. These facilities may include pilot plants, and such research facilities may be located at a site separate from the production plant(s).

World Chemical Production

The scale of chemical manufacturing tends to be organized from largest in volume (petrochemicals and commodity chemicals), to specialty chemicals, and the smallest, fine chemicals.

The petrochemical and commodity chemical manufacturing units are on the whole single product continuous processing plants. Not all petrochemical or commodity chemical materials are made in one single location, but groups of related materials often are to induce industrial symbiosis as well as material, energy and utility efficiency and other economies of scale.

Distillation columns.

Those chemicals made on the largest of scales are made in a few manufacturing locations around the world, for example in Texas and Louisiana along the Gulf Coast of the United States, on Teesside in the Northeast of England in the United Kingdom, and in Rotterdam in the Netherlands. The large scale manufacturing locations often have clusters of manufacturing units that share utilities and large scale infrastructure such as power stations, port facilities, road and rail terminals. To demonstrate the clustering and integration mentioned above, some 50% of the United Kingdom's petrochemical and commodity chemicals are produced by the Northeast of England Process Industry Cluster on Teesside.

Specialty chemical and fine chemical manufacturing are mostly made in discrete batch processes. These manufacturers are often found in similar locations but in many cases they are to be found in multi sector business parks.

Continents and Countries

In the U.S. there are 170 major chemical companies. They operate internationally with more than 2,800 facilities outside the U.S. and 1,700 foreign subsidiaries or affiliates operating. The U.S. chemical output is $750 billion a year. The U.S. industry records large trade surpluses and employs more than a million people in the United States alone. The chemical industry is also the second largest consumer of energy in manufacturing and spends over $5 billion annually on pollution abatement.

In Europe the chemical, plastics and rubber sectors are among the largest industrial sectors. Together they generate about 3.2 million jobs in more than 60,000 companies. Since 2000 the chemical sector alone has represented 2/3 of the entire manufacturing trade surplus of the EU.

in 2012 The chemical sector accounted for 12% of the EU manufacturing industry's added value. Europe remains world's biggest chemical trading region with 43% of the world's exports and 37%of the world's imports, although the latest data shows that Asia is catching up with 34% of the exports and 37% of imports. Even so, Europe still has a trading surplus with all regions of the world except Japan and China where in 2011 there was a chemical trade balance. Europe's trade surplus with the rest of the world today amounts to 41.7 billion Euros.

Over the 20 years between 1991 and 2011 the European Chemical industry saw its sales increase 295 billion Euros to 539 billion Euros a picture of constant growth. Despite this the European industry's share of the world chemical market has fallen from 36% to 20%. This has resulted from the huge increase production and sales in the emerging markets like India and China. The data suggest that 95% of this impact is from China alone. In 2012 the data from the European Chemical Industry Council (CEFIC)shows that 5 European countries account for 71% of the EU's chemicals sales. These are Germany, France, United Kingdom, Italy and the Netherlands.

The chemical industry has shown rapid growth for more than fifty years. The fastest-growing areas have involved the manufacture of synthetic organic polymers used as plastics, fibers and elastomers. Historically and presently the chemical industry has been concentrated in three areas of the world, Western Europe, North America and Japan (the Triad). The European Community remains the largest producer area followed by the US and Japan.

The traditional dominance of chemical production by the Triad countries is being challenged by changes in feedstock availability and price, labor cost, energy cost, differential rates of economic growth and environmental pressures. Instrumental in the changing structure of the global chemical industry has been the growth in China, India, Korea, the Middle East, South East Asia, Nigeria, and Brazil.

Just as companies emerge as the main producers of the chemical industry, we can also look on a more global scale to how industrialized countries rank, with regards to the billions of dollars worth of production a country or region could export. Though the business of chemistry is worldwide in scope, the bulk of the world's $3.7 trillion chemical output is accounted for by only a handful of industrialized nations. The United States alone produced $689 billion, 18.6 percent of the total world chemical output in 2008.

Global Chemical Shipments by Country/ Region (billions of dollars)	1998	1999	2000	2001	2002	2003	2004	2005	2006	**2008**	**2009**
United States of America	416.7	420.3	449.2	438.4	462.5	487.7	540.9	610.9	657.7	664.1	689.3
Canada	21.1	21.8	25.0	24.8	25.8	30.5	36.2	40.2	43.7	45.4	47.4
Mexico	19.1	21.0	23.8	24.4	24.3	23.5	25.6	29.2	32.0	33.4	37.8
North America	**456.9**	463.1	498.0	487.6	512.6	541.7	602.7	680.3	733.4	742.8	774.6
Brazil	46.5	40.0	45.7	41.5	39.6	47.4	60.2	71.1	82.8	96.4	126.7
Other	59.2	58.1	60.8	63.4	58.6	62.9	69.9	77.2	84.6	89.5	102.1
Latin America	**105.7**	**98.1**	**106.5**	**104.9**	98.2	110.3	130.0	148.3	167.4	185.9	228.8
France	79.1	78.5	76.5	76.8	80.5	99.6	111.1	117.5	121.3	138.4	158.9
Germany	124.9	123.2	118.9	116.1	120.1	148.1	168.6	178.6	192.5	229.5	263.2
Italy	63.9	64.6	59.5	58.6	64.5	75.8	86.6	89.8	95.3	105.9	122.9
United Kingdom	70.3	70.1	66.8	66.4	69.9	77.3	91.3	95.2	107.8	118.2	123.4
Belgium	27.1	27.0	27.5	27.1	28.7	36.1	41.8	43.5	46.9	51.6	62.6
Ireland	16.9	20.1	22.6	22.9	29.1	32.3	33.9	34.9	37.5	46.0	54.8
Netherlands	29.7	29.4	31.3	30.6	32.2	40.1	49.0	52.7	59.2	67.9	81.7
Spain	31.0	30.8	30.8	31.9	33.4	42.0	48.9	52.7	56.7	63.7	74.8
Sweden	11.1	11.4	11.2	11.0	12.5	15.9	18.2	19.3	21.2	21.2	22.6
Switzerland	22.1	22.2	19.4	21.1	25.5	30.3	33.8	35.4	37.8	42.7	53.1
Other	27.1	26.8	25.9	26.4	27.9	33.5	38.6	42.9	46.2	50.3	58.9
Western Europe	503.1	504.0	490.4	488.8	524.4	630.9	721.9	762.7	822.4	935.4	1,076.8
Russia	23.8	24.6	27.4	29.1	30.3	33.4	37.5	40.9	53.1	63.0	77.6
Other	22.3	20.3	21.9	23.4	25.3	31.4	39.6	46.2	55.0	68.4	87.5
Central/ Eastern Europe	**46.1**	**44.9**	49.3	52.5	55.6	64.8	77.1	87.1	108.0	131.3	165.1
Africa & Middle East	52.7	53.2	59.2	57.4	60.4	73.0	86.4	99.3	109.6	124.2	160.4
Japan	193.8	220.4	239.7	208.3	197.2	218.8	243.6	251.3	248.5	245.4	298.0
Asia-Pacific excluding Japan	215.2	241.9	276.1	271.5	300.5	369.1	463.9	567.5	668.8	795.5	993.2
China	80.9	87.8	103.6	111.0	126.5	159.9	205.0	269.0	331.4	406.4	549.4
India	30.7	35.3	35.3	32.5	33.5	40.8	53.3	63.6	72.5	91.1	98.2

Australia	11.3	12.1	11.2	10.8	11.3	14.9	17.0	18.7	19.1	22.8	27.1
Korea	39.3	45.5	56.3	50.4	54.9	64.4	78.7	91.9	103.4	116.7	133.2
Singapore	6.3	8.5	9.5	9.4	12.5	16.1	20.0	22.0	25.8	28.9	31.6
Taiwan	21.9	23.7	29.2	26.8	28.4	34.3	44.5	49.5	53.8	57.4	62.9
Other Asia/ Pacific	24.8	29.1	30.9	30.8	33.3	38.8	45.5	52.9	62.9	72.2	90.8
Asia/Pacific	**409.0**	**462.3**	**515.7**	**479.7**	**497.7**	**587.8**	**707.5**	**818.8**	**917.3**	**1041.0**	**1291.2**
Total world shipments	1573.5	1625.5	1719.0	1670.9	1748.8	2008.5	2325.6	2596.4	2858.1	3160.7	3696.8

CHEMICAL LEASING

In the chemical industry, chemical leasing is a business model in which the chemical company supplies a substance for a specific service, but retains ownership of the chemical. It is intended to shift the focus from increasing sales volume of chemicals towards a value added approach. It may lead to more efficient use of chemicals, and to quality, environmental, and economic benefits.

Definitions

The producer mainly sells the functions performed by the chemical and the functional units, such as the number of pieces painted, are the main basis for payment. It is a business model in which a customer engages with a service provider in a strategic, long-term contract to supply and manage the customer's chemical and related services. This model has been practiced by the paint industry for decades. For example, the painting of roads or sidewalks are often charged on the basis of the number of kilometers to be covered, rather than on the liters of paint consumed. This concept is referred to by several different names, such as "eco-efficient services", "product-service systems", "chemical servicizing", "chemical management services", and the "product service system".

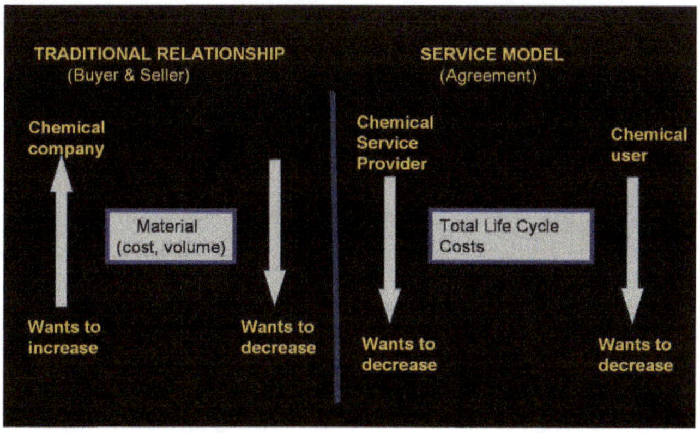

Differences between traditional buyer-seller relationship and service based model of chemical leasing.

Chemical leasing promotes the sustainable management of chemicals. By shifting the focus from increasing the sales volume of chemicals towards a more value added approach, it is an illustration of extended producer responsibility. The chemical company supplies a substance for a specific service, such as powder coating, dyeing or painting, but retains ownership over the chemical, and also advises the user on its best use. The producer owns the machinery, and maintains responsibility over the manufacturing process. As a result, it is in the interest of all parties to use the substances as efficiently as possible.

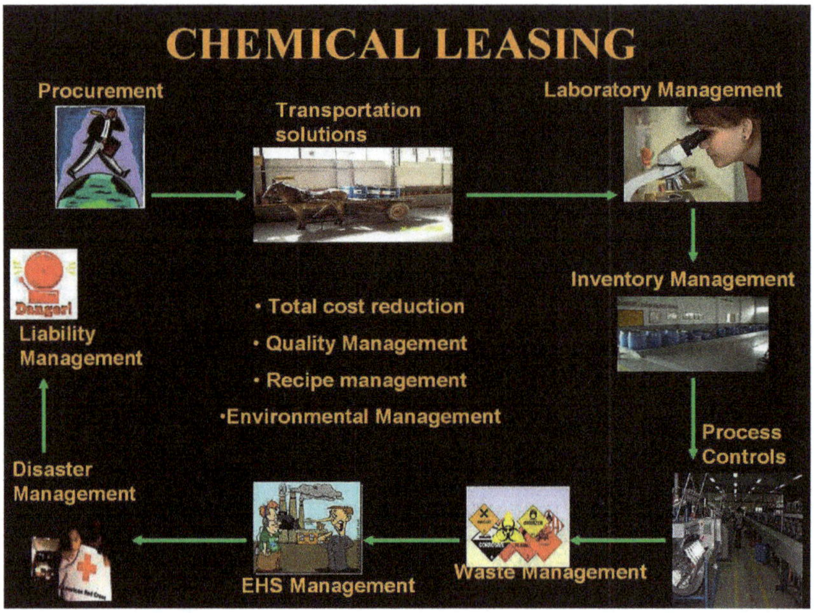

The scope of chemical leasing includes the entire life cycle of chemicals used in a factory.

Globalization

At the Earth Summit 2002 on sustainable development, the international community agreed on the goal of ensuring that, by the year 2020, chemicals will be produced and used in ways that minimize significant adverse impacts on human health and the environment. As a consequence the United Nations Environmental Programme (UNEP) decided to develop of the Strategic Approach to International Chemicals Management (SAICM) in 2003. In 2006 at the International Conference on Chemicals Management (ICCM) in Dubai, the signing countries committed themselves to promote the sound management of chemicals and hazardous wastes at all levels.

The Austrian government has played an important role in promoting chemicals management based on resource efficiency and precaution. When Austria held the Presidency of the European Union during the first half of 2006, chemicals policy was on top of the environmental agenda. Austria continues to promote chemical leasing.

The Federal Environment Agency (UBA) of Germany proposed to promote chemical leasing in its "Sustainable Chemicals" paper.

Chemical Management Services

Integrated chemical services, also called total chemical management or CMS, include the development of chemicals according to customer requirements, the provision of logistics solutions, administration, process development, chemical application, and waste management. The majority of chemical management services in Europe are large international chemical companies such as Dow, Castrol, Quaker, PPG, and BASF that were able to broaden their market share or become the sole providers of CMS by extending the range of their chemicals and adding information management and technical services to their chemicals business.

CMS customers in Europe and America are primarily large, multinational Fortune 500 companies in the automotive, metalworking, aerospace, and electronics sectors. Automotive companies such as Volkswagen, Daimler Chrysler, Volvo, Ford, Toyota, and General Motors have been outsourcing the management of their paint shops in Europe and some of these facilities are now also moving toward total outsourcing of chemical management. Prominent companies in the semiconductor industry, such as STMicroelectronics, Micron Technologies, and Motorola, have also been using chemical management services for many years. New areas of CMS growth in Europe include the pulp and paper industry and the life sciences industry.

UNIDO

In 2004, the United Nations Industrial Development Organization (UNIDO) and the Austrian Ministry of Environment decided to jointly support chemical leasing through a number of global projects. In partnership with National Cleaner Production Centers, UNIDO has implemented chemical leasing projects in Austria, Egypt, Mexico and Russia. PERO Innovative Services GmbH together with SAFECHEM Europe GmbH, a subsidiary of The Dow Chemical Company, have supported Automobiltechnik Blau in metal cleaning to use cost efficient machines, lower the energy consumption and chemical usage.

Akzo Nobel Powder Coatings S.A.E has supported the chemical leasing of powder coating in Egypt to ABB ARAB. Environmental benefits are said to include recycling of powder waste, compliance with environmental regulations, and enhancement of supply chain management. General Motors Egypt has tied up with the Dr. Badawi Chemical Work to lease hydrocarbon solvents used in metal finishing. Elsewedy Electric, the Egyptian market leader in hot dip galvanization, has partnered with Zinc Misr to manage the use of flux chemicals and reduce waste discharges.

Mepla–Alfit and Tiefenbacher GmbH in Austria are cooperating in the chemical leasing of de-lacquering agents and paint strippers. IKEA and the Huntsman Corporation are cooperating in India to secure the colouring process in textiles.

The sugar mill Fideicomiso Ingenio San Cristobal in Mexico has implemented chemical leasing for the machine lubrication with the help of Shutz Oil and SUMAT. Henkel-ERA in Russia is using the services of ERG for industrial waste water treatment.

A Joint Declaration of Intent on Chemical Leasing was signed between UNIDO, Austria, Germany and Switzerland in November 2016.

REACH

The reversal of the burden of proof is a key component of the Registration, Evaluation, Authorization and Restriction of Chemicals (REACH) system leading to a "no data – no market" concept, obliging producers or importers of substances to deliver documentation regarding properties of chemicals and possible risks during their application as a pre-condition for market access. The OECD Conference in Vienna in 2003 Nov. reiterated that, "... The new EU chemicals policy (REACH) will require a new relationship between provider and user ...." According to Thomas Jakl, Chairman of the European Chemical Agency (ECHA), chemical leasing paves the way to comply with REACH obligations.

Chemical leasing and REACH share the same philosophy of ensuring compliance with a duty of care (REACH recital 16), as a tool to demonstrate adequate control (REACH para 60). They are mutually supportive in developing rules for sharing costs, and ensuring that chemicals are handled properly. Both are involve several different stages of the supply chain. There is a strong effort by the Austrian and German Governments to bring chemical leasing within the purview of EU Chemicals policy and regulations.

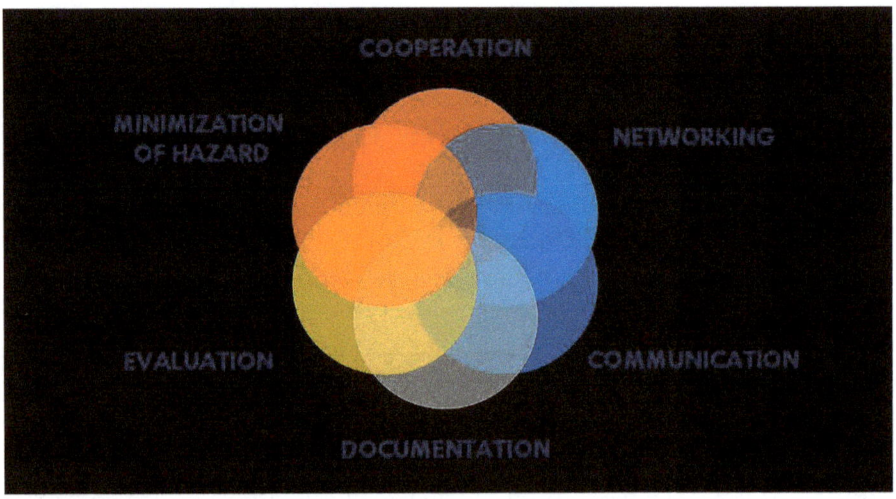

Synergies between chemical leasing and EU REACH legislation.

Project Phases

Chemical leasing projects are divided into planning, implementation, evaluation and dissemination stages, based on a Deming Cycle. The planning stage consists of a preparatory phase, a process optimization phase and a design phase. In this stage, discussions around the leasing model, its cost implications versus quality and environmental benefits, commercial terms, and conditions begin. A baseline audit is performed, and a report presented to the factory management. This audit outlines the potential for improvements and forms the basis of defining the key performance indicators (KPIs). The resources needed to fulfil improvements are also defined.

The implementation stage starts with the signing of a chemical leasing agreement that defines the scope and conditions, unit of payment, KPIs, roles, and responsibilities. The chemical company supervises the chemical process, transporting and managing the inventory, laboratory management, improving process controls, record keeping, and training workers. Periodic checks and inspections are carried out independently to verify that the implementation is proceeding on expected lines. At the end of the implementation phase, progress is evaluated, often by an external party to secure objectivity. Finally, any project benefits are quantified and learning is documented, to provide input for future projects.

REFERENCES

1. Derry, Thomas Kingston; Williams, Trevor I. (1993). A Short History of Technology: From the Earliest Times to A.D. 1900. New York: Dover.
2. Kiefer, David M. (2001). "Sulfuric Acid: Pumping Up the Volume". American Chemical Society. Retrieved 2008-04-21.
3. "The Chemical Industries In The UK". American Chemical Society. Retrieved 2013-04-21.
4. Aftalion, Fred (1991). A History of the International Chemical Industry. Philadelphia: University of Pennsylvania Press. pp. 11–13. ISBN 0-8122-1297-5.
5. Aftalion, Fred (1991). A History of the International Chemical Industry. Philadelphia: University of Pennsylvania Press. pp. 14–16. ISBN 0-8122-1297-5.
6. Aftalion 1991, p. 104, Chandler 2004, p. 475
7. "Electrolytic Production of Bromine - National Historic Chemical Landmark - American Chemical Society". American Chemical Society. Retrieved 2016-10-10.
8. UK Patent office (1857). Patents for inventions. UK Patent office. p. 255.
9. Jeannifer Filly Sumayku, Unilever: Providing Enjoyable and Meaningful Life to Customers, The President Post, 22 March 2010
10. Centi, Gabriele (January 2012). "Resources and energy efficiency: the winning strategy for chemical industry" (PDF). La Chimica l'Industria. Società Chimica Italiana (1): 64–67.
11. "PTFE Properties". Fluorotherm Polymers, Inc. Retrieved 31 October 2014.
12. "Sectors of Chemical Industry". Technofunc. Retrieved 16 September 2013.
13. Global Specialty Chemicals (PDF) (Report). =Marketline. May 2012. Retrieved 16 September 2012.
14. Global Speciality Chemicals (PDF) (Report). MarketLine. May 2012.
15. Tullo, Alexander H. "C&EN's Global Top 50 | July 25, 2016 Issue - Vol. 94 Issue 30 | Chemical & Engineering News". cen.acs.org. Retrieved 2016-10-10.
16. "Chemical and Agrochemical Enterprise Quality Management Software". Sparta Systems, Inc. Retrieved 20 March 2015.
17. "Facts and Figures 2012:The European chemicals industry in a worldwide perspective" (PDF). CEFIC. Retrieved 5 August 2013.

18. Higgins, Stan (April 2013). "European Chemicals Industry:A review" (PDF). Chemical News. pp. 18–20.
19. "Facts and Figures 2012:The European chemicals industry in a worldwide perspective" (PDF). CEFIC: 6. Retrieved 5 August 2013.
20. "Facts and Figures 2012:The European chemicals industry in a worldwide perspective" (PDF). CEFIC: 7. Retrieved 5 August 2013.
21. "Global Business of Chemistry". Retrieved 26 February 2016.
22. 'World Of Chemicals' is a chemical portal - a networking community for the chemical fraternity. http://www.worldofchemicals.com/
23. Fred Aftalion A History of the International Chemical Industry. University of Pennsylvania Press. 1991. online version
24. E. N. Brandt. Growth Company: Dow Chemical's First Century. Michigan State University Press. xxii+ 650 pp. Appendices, Select bibliography and index. ISBN 0-87013-426-4. online review
25. Alfred D. Chandler. Shaping the Industrial Century: The Remarkable Story of the Evolution of the Modern Chemical and Pharmaceutical Industries. Harvard University Press, 2005. 366 pp. ISBN 0-674-01720-X.
26. Micheal McCoy, et al., "Facts & Figures of the Chemical Industry", Chemical & Engineering News, 84(28), July 10, 2006, pp. 35–72.
27. Shreve, R. Norris, and Joseph A. Brink Jr. The Chemical Process Industries (4th ed. New York: McGraw Hill, 1977)
28. Woytinsky, W. S., and E. S. Woytinsky. World Population and Production Trends and Outlooks (1953) pp 1176–1205;
29. Chemical refinery resources http://www.ccc-group.com/chemicals/solutions/industrial-1
30. United Nations Industrial Development Organization
31. Chemical Leasing Goes Global – Selling Services Instead of Barrels: A Win-Win Business Model for Environment and Industry, Eds Thomas Jakl and Petra Schwager, (Wien : Springer – Verlag, 2008).
32. M Stoughton, T Votta, Implementing service-based chemical procurement: Lessons and results. Journal of Cleaner Production 11(8) (2003) 839– 849.
33. Anurag Priyadarshi, The Colourist, 1 (4) (2009) 8.
34. Chemical Leasing Unburdens Environment – UBA presents sustainable chemicals concept, UBA Press Release 14 (Umwelt Bundes Amt, Germany, 2009).
35. Oksana Mont, P Singhal, Z Fadeeva, Chemical Management Services in Sweden and Europe, J. Indl. Eco., 10 (1) (2006) 279.
36. Case Study: General Motors, Navistar, Chrysler Neon, Ford Taurus, Delta Air Lines, Chemical Strategies Partnership, San Francisco, USA (2000) (available at www.chemicalstrategies.org).

37. C Weigel, Service strategies for value enhancement in the chemical industry. Paper read at OECD conference Experiences and Perspectives of Service-Oriented Strategies in the Chemicals Industry and in Related Areas, 13–14 November (2003) Vienna.

38. IKEA's Sustainability Report 2008

39. P Schwager, Chemical Leasing: Interregional Promotion and Implementation of Closing-the-loops Cooperation and Business Models in the Chemical Industry (Vienna: United Nations Industrial Development Organization, 2008).

40. P Schwager, F Moser, The application of Chemical Leasing business model in Mexico, ESPR, 13 (2) (2006) 131.

41. "Joint Declaration of Intent on Chemical Leasing signed between UNIDO, Austria, Germany and Switzerland".

42. Chemical Leasing Goes Global – Selling Services Instead of Barrels: A Win-Win Business Model for Environment and Industry, Eds T Jakl and P Schwager,

43. Anurag Priyadarshi, Chemical Leasing in Colour and Environmental Management – theoretical framework, implementation methodology and future outlook, paper read at UNIDO – GCPC Chemical Leasing workshop, 2009.

44. "Global Chemical Leasing Award". Retrieved 28 September 2012.

Unit Operations

Chapter Contents

Unit operation is the operation of transforming the material without changing its chemical properties. It does not involve any reactions. Unit operation involves all types of operations carried out in chemical and process industries which do not involve any chemical reaction. Hence the chemical properties of the substance do not change however the physical properties may change. Examples are size reduction process, Separation operations and heat transfer operations

INTRODUCTION

In chemical engineering and related fields, a unit operation is a basic step in a process. Unit operations involve a physical change or chemical transformation such as separation, crystallization, evaporation, filtration, polymerization, isomerization, and other reactions. For example, in milk processing, homogenization, pasteurization, chilling, and packaging are each unit operations which are connected to create the overall process. A process may require many unit operations to obtain the desired product from the starting materials, or feedstocks.

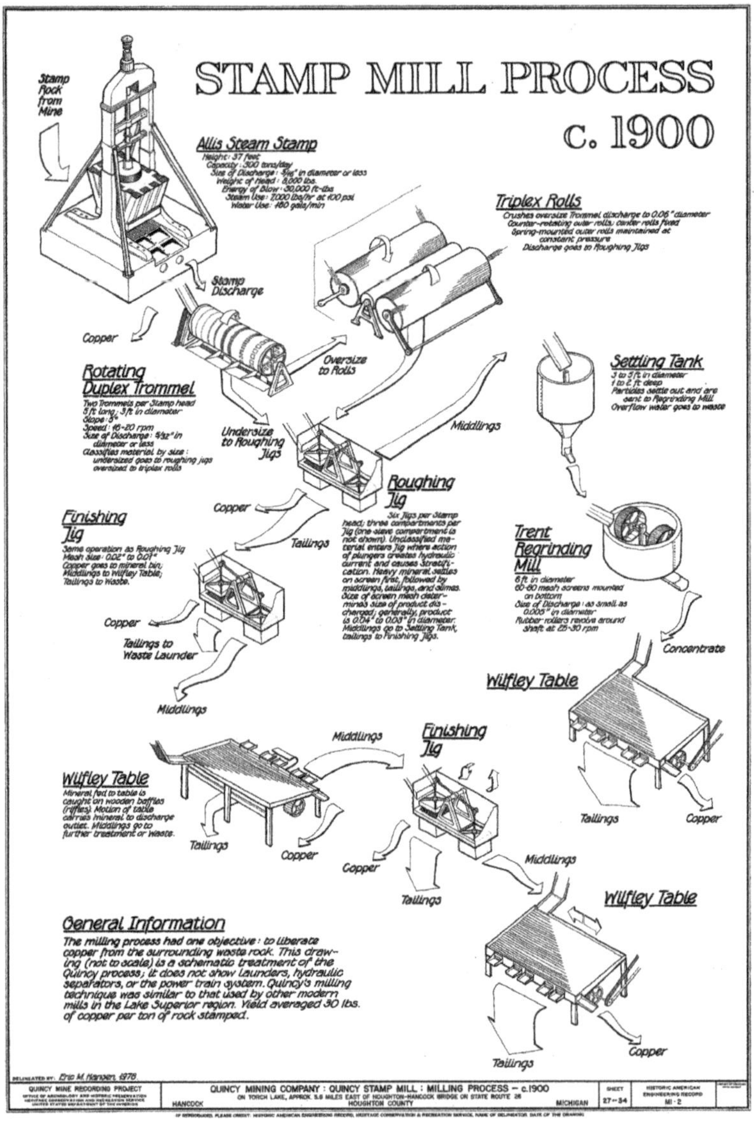

An ore extraction process broken into its constituent unit operations (Quincy Mine, Hancock, MI ca. 1900).

HISTORY

Historically, the different chemical industries were regarded as different industrial processes and with different principles. Arthur Dehon Little propounded the concept of "unit operations" to explain industrial chemistry processes in 1916. In 1923, William H. Walker, Warren K. Lewis and William H. McAdams wrote the book *The Principles of Chemical Engineering* and explained that the variety of chemical industries have processes which follow the same physical laws. They summed up these similar processes into unit operations. Each unit operation follows the same physical laws and may be used in all relevant chemical industries. For instance, the same engineering is required to design a mixer for either napalm or porridge, even if the use, market or manufacturers are very different. The unit operations form the fundamental principles of chemical engineering.

Chemical engineering unit operations consist of five classes:

- Fluid flow processes, including fluids transportation, filtration, and solids fluidization.
- Heat transfer processes, including evaporation and heat exchange.
- Mass transfer processes, including gas absorption, distillation, extraction, adsorption, and drying.
- Thermodynamic processes, including gas liquefaction, and refrigeration.
- Mechanical processes, including solids transportation, crushing and pulverization, and screening and sieving.

Chemical engineering unit operations also fall in the following categories which involve elements from more than one class:

- Combination (mixing)
- Separation (distillation, crystallization)
- Reaction (chemical reaction)

Furthermore, there are some unit operations which combine even these categories, such as reactive distillation and stirred tank reactors. A "pure" unit operation is a physical transport process, while a mixed chemical/physical process requires modeling both the physical transport, such as diffusion, *and* the chemical reaction. This is usually necessary for designing catalytic reactions, and is considered a separate discipline, termed chemical reaction engineering.

Chemical engineering unit operations and chemical engineering unit processing form the main principles of all kinds of chemical industries and are the foundation of designs of chemical plants, factories, and equipment used.

In general, unit operations are designed by writing down the balances for the transported quantity for each elementary component (which may be infinitesimal) in the form of equations, and solving the equations for the design parameters, then selecting an optimal solution out of the several possible and then designing the physical equipment. For instance, distillation in a plate column is analyzed by writing down the mass balances for each plate, wherein the known vapor-liquid equilibrium and efficiency, drip out and drip in comprise the total mass flows, with a sub-flow for each component. Combining a stack of these gives the system of equations for the whole column. There is a range of solutions, because a higher reflux ratio enables fewer plates, and vice versa. The engineer must then find the optimal solution with respect to acceptable volume holdup, column height and cost of construction.

MOMENTUM TRANSFER

In particle physics, wave mechanics and optics, momentum transfer is the amount of momentum that one particle gives to another particle.

In the simplest example of scattering of two colliding particles with initial momenta $\vec{p}_{i1}, \vec{p}_{i2}$, resulting in final momenta $\vec{p}_{f1}, \vec{p}_{f2}$, the momentum transfer is given by

$$\vec{q} = \vec{p}_{i1} - \vec{p}_{f1} = \vec{p}_{f2} - \vec{p}_{i2}$$

where the last identity expresses momentum conservation. Momentum transfer is an important quantity because $\Delta x = \hbar/|q|$ is a better measure for the typical distance resolution of the reaction than the momenta themselves.

Wave Mechanics and Optics

A wave has a momentum $p = \hbar k$ and is a vectorial quantity. The difference of the momentum of the scattered wave to the incident wave is called *momentum transfer*. The wave number k is the absolute of the wave vector $k = q/\hbar$ and is related to the wavelength $k = 2\pi/\lambda$. Often, momentum transfer is given in wavenumber units in reciprocal length $Q = k_f - k_i$

Diffraction

The momentum transfer plays an important role in the evaluation of neutron, X-ray and electron diffraction for the investigation of condensed matter. Bragg diffraction occurs on the atomic crystal lattice, conserves the wave energy and thus is called elastic scattering, where the wave numbers final and incident particles, k_f and k_i, respectively, are equal and just the direction changes by a reciprocal lattice vector $G = Q = k_f - k_i$ with the relation to the lattice spacing $G = 2\pi/d$. As momentum is conserved, the transfer of momentum occurs to crystal momentum.

The presentation in Q-space is generic and does not depend on the type of radiation and wavelength used but only on the sample system, which allows to compare results obtained from many different methods. Some established communities such as powder diffraction employ the diffraction angle 2θ as the independent variable, which worked fine in the early years when only a few characteristic wavelengths such as Cu-Kα were available. The relationship to Q-space is

$$Q = \frac{4\pi \sin(\theta)}{\lambda}$$

and basically states that larger 2θ corresponds to larger Q.

Momentum-Transfer Cross Section

In physics, and especially scattering theory, the momentum-transfer cross section (sometimes known as the momentum-*transport* cross section) is an effective scattering cross section useful for describing the average momentum transferred from a particle when it collides with a target. Essentially, it contains all the information about a scattering process necessary for calculating average momentum transfers but ignores other details about the scattering angle.

The momentum-transfer cross section σ_{tr} is defined in terms of an (azimuthally symmetric and momentum independent) differential cross section $\frac{\mathrm{d}\sigma}{\mathrm{d}\Omega}(\theta)$ by

$$\sigma_{\text{tr}} = \int (1-\cos\theta)\frac{\mathrm{d}\sigma}{\mathrm{d}\Omega}(\theta)\mathrm{d}\Omega$$

$$= \int\int (1-\cos\theta)\frac{\mathrm{d}\sigma}{\mathrm{d}\Omega}(\theta)\sin\theta\mathrm{d}\theta\mathrm{d}\phi.$$

The momentum-transfer cross section can be written in terms of the phase shifts from a partial wave analysis as

$$\sigma_{\text{tr}} = \frac{4\pi}{k^2}\sum_{l=0}^{\infty}(l+1)\sin^2[\delta_{l+1}(k)-\delta_l(k)].$$

Explanation

The factor of $1-\cos\theta$ arises as follows. Let the incoming particle be traveling along the z-axis with vector momentum

$$\vec{p}_{\text{in}} = q\hat{z}.$$

Suppose the particle scatters off the target with polar angle θ and azimuthal angle

$$\vec{p}_{\text{out}} = q'\cos\theta\hat{z} + q'\sin\theta\cos\phi\hat{x} + q'\sin\theta\sin\phi\hat{y}.$$

For collision to much heavier target than striking particle (ex: electron incident on the atom or ion), $q' \simeq q$ so

$$\vec{p}_{\text{out}} \simeq q\cos\theta\hat{z} + q\sin\theta\cos\phi\hat{x} + q\sin\theta\sin\phi\hat{y}$$

By conservation of momentum, the target has acquired momentum

$$\Delta\vec{p} = \vec{p}_{\text{in}} - \vec{p}_{\text{out}} = q(1-\cos\theta)\hat{z} - q\sin\theta\cos\phi\hat{x} - q\sin\theta\sin\phi\hat{y}.$$

Now, if many particles scatter off the target, and the target is assumed to have azimuthal symmetry, then the radial (x and y) components of the transferred momentum will average to zero. The average momentum transfer will be just $q(1-\cos\theta)\hat{z}$. If we do the full averaging over all possible scattering events, we get

$$\begin{aligned}\Delta\vec{p}_{\text{avg}} &= \langle\Delta\vec{p}\rangle_\Omega \\ &= \sigma_{\text{tot}}^{-1}\int \Delta\vec{p}(\theta,\phi)\frac{\mathrm{d}\sigma}{\mathrm{d}\Omega}(\theta)\mathrm{d}\Omega \\ &= \sigma_{\text{tot}}^{-1}\int [q(1-\cos\theta)\hat{z} - q\sin\theta\cos\phi\hat{x} - q\sin\theta\sin\phi\hat{y}]\frac{\mathrm{d}\sigma}{\mathrm{d}\Omega}(\theta)\mathrm{d}\Omega \\ &= q\hat{z}\sigma_{\text{tot}}^{-1}\int (1-\cos\theta)\frac{\mathrm{d}\sigma}{\mathrm{d}\Omega}(\theta)\mathrm{d}\Omega \\ &= q\hat{z}\sigma_{\text{tr}}/\sigma_{\text{tot}}\end{aligned}$$

where the total cross section is

$$\sigma_{\text{tot}} = \int \frac{\mathrm{d}\sigma}{\mathrm{d}\Omega}(\theta)\mathrm{d}\Omega.$$

Here, the averaging is done by using expected value calculation. Therefore, for a given total cross section, one does not need to compute new integrals for every possible momentum in order to determine the average momentum transferred to a target. One just needs to compute σ_{tr}.

Application

This concept is used in calculating charge radius of nuclei such as proton and deuteron by electron scattering experiments.

To this purpose a useful quantity called the scattering vector q having the dimension of inverse length is defined as a function of energy E and scattering angle θ:

$$q = \frac{\frac{2E}{\hbar c}\sin(\theta/2)}{[1+\frac{2E}{Mc^2}\sin^2(\theta/2)]^{1/2}}$$

HEAT TRANSFER

Heat transfer is the exchange of thermal energy between physical systems. The rate of heat transfer is dependent on the temperatures of the systems and the properties of the intervening medium through which the heat is transferred. The three fundamental modes of heat transfer are *conduction*, *convection* and *radiation*. Heat transfer, the flow of energy in the form of heat, is a process by which a system's internal energy is changed, hence is of vital use in applications of the First Law of Thermodynamics. Conduction is also known as diffusion, not to be confused with diffusion related to the mixing of constituents of a fluid.

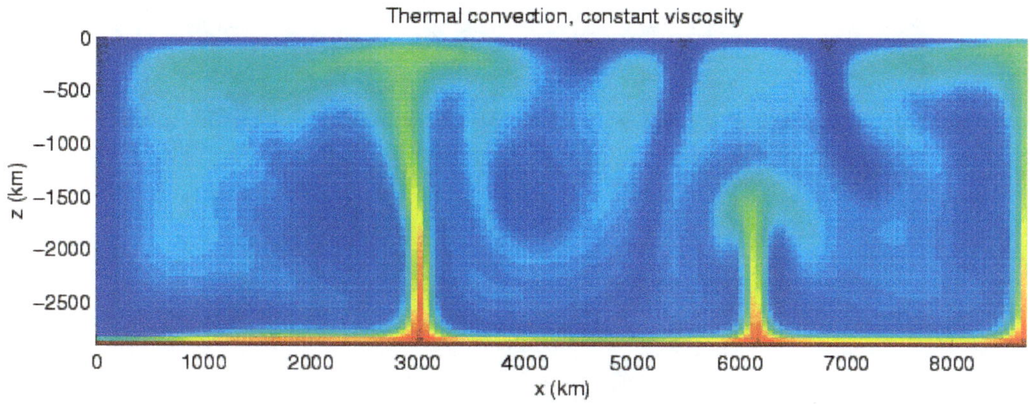

This figure shows a calculation for thermal convection in the Earth's mantle. Colors closer to red are hot areas and colors closer to blue are cold areas. A hot, less-dense lower boundary layer sends plumes of hot material upwards, and likewise, cold material from the top moves downwards.

The direction of heat transfer is from a region of high temperature to another region of lower temperature, and is governed by the Second Law of Thermodynamics. Heat transfer changes the internal energy of the systems from which and to which the energy is transferred. Heat transfer will occur in a direction that increases the entropy of the collection of systems.

Heat transfer ceases when thermal equilibrium is reached, at which point all involved bodies and the surroundings reach the same temperature. Thermal expansion is the tendency of matter to change in volume in response to a change in temperature.

Overview

Heat is defined in physics as the transfer of thermal energy across a well-defined boundary around a thermodynamic system. The thermodynamic free energy is the amount of work that a thermodynamic system can perform. Enthalpy is a thermodynamic potential, designated by the letter "H", that is the sum of the internal energy of the system (U) plus the product of pressure (P) and volume (V). Joule is a unit to quantify energy, work, or the amount of heat.

Heat transfer is a process function (or path function), as opposed to functions of state; therefore, the amount of heat transferred in a thermodynamic process that changes the state of a system depends on how that process occurs, not only the net difference between the initial and final states of the process.

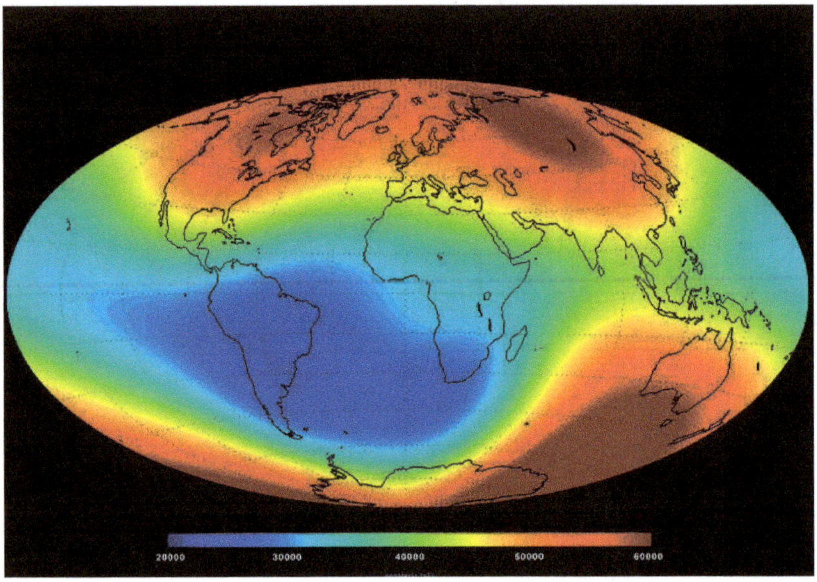

Earth's longwave thermal radiation intensity, from clouds, atmosphere and surface.

Thermodynamic and mechanical heat transfer is calculated with the heat transfer coefficient, the proportionality between the heat flux and the thermodynamic driving force for the flow of heat. Heat flux is a quantitative, vectorial representation of heat-flow through a surface.

In engineering contexts, the term *heat* is taken as synonymous to thermal energy. This usage has its origin in the historical interpretation of heat as a fluid (*caloric*) that can be transferred by various causes, and that is also common in the language of laymen and everyday life.

The transport equations for thermal energy (Fourier's law), mechanical momentum (Newton's law for fluids), and mass transfer (Fick's laws of diffusion) are similar, and analogies among these three transport processes have been developed to facilitate prediction of conversion from any one to the others.

Thermal engineering concerns the generation, use, conversion, and exchange of heat transfer. As such, heat transfer is involved in almost every sector of the economy. Heat transfer is classified into various mechanisms, such as thermal conduction, thermal convection, thermal radiation, and transfer of energy by phase changes.

Mechanisms

The fundamental modes of heat transfer are:

Advection

Advection is the transport mechanism of a fluid from one location to another, and is dependent on motion and momentum of that fluid.

Conduction or Diffusion

The transfer of energy between objects that are in physical contact. Thermal conductivity is the property of a material to conduct heat and evaluated primarily in terms of Fourier's Law for heat conduction.

Convection

The transfer of energy between an object and its environment, due to fluid motion. The average temperature is a reference for evaluating properties related to convective heat transfer.

Radiation

The transfer of energy by the emission of electromagnetic radiation.

Advection

By transferring matter, energy—including thermal energy—is moved by the physical transfer of a hot or cold object from one place to another. This can be as simple as placing hot water in a bottle and heating a bed, or the movement of an iceberg in changing ocean currents. A practical example is thermal hydraulics. This can be described by the formula:

$$Q = v\rho c_p \Delta T$$

where Q is heat flux (W/m^2), ρ is density (kg/m^3), C_p is heat capacity at constant pressure (J/kg·K), ΔT is the change in temperature (K), v is velocity (m/s).

Conduction

On a microscopic scale, heat conduction occurs as hot, rapidly moving or vibrating atoms and molecules interact with neighboring atoms and molecules, transferring some of their energy (heat) to these neighboring particles. In other words, heat is transferred by conduction when adjacent atoms vibrate against one another, or as electrons move from one atom to another. Conduction is the most significant means of heat transfer within a solid or between solid objects in thermal contact. Fluids—especially gases—are less conductive. Thermal contact conductance is the study of heat conduction between solid bodies in contact. The process of heat transfer from one place to another place without the movement of particles is called conduction Example:Heat transfer through Metal rods. *Steady state conduction* is a form of conduction that happens when the temperature difference driving the conduction is constant, so that after an equilibration time, the spatial distribution of temperatures in the conducting object does not change any further. In steady state conduction, the amount of heat entering a section is equal to amount of heat coming out.

Transient conduction occurs when the temperature within an object changes as a function of time. Analysis of transient systems is more complex and often calls for the application of approximation theories or numerical analysis by computer.

Convection

The flow of fluid may be forced by external processes, or sometimes (in gravitational fields) by buoyancy forces caused when thermal energy expands the fluid (for example in a fire plume), thus influencing its own transfer. The latter process is often called "natural convection". All convective processes also move heat partly by diffusion, as well. Another form of convection is forced convection. In this case the fluid is forced to flow by use of a pump, fan or other mechanical means.

Convective heat transfer, or convection, is the transfer of heat from one place to another by the movement of fluids, a process that is essentially the transfer of heat via mass transfer. Bulk motion of fluid enhances heat transfer in many physical situations, such as (for example) between a solid surface and the fluid. Convection is usually the dominant form of heat transfer in liquids and gases. Although sometimes discussed as a third method of heat transfer, convection is usually used to describe the combined effects of heat conduction within the fluid (diffusion) and heat transference by bulk fluid flow streaming. The process of transport by fluid streaming is known as advection, but pure advection is a term that is generally associated only with mass transport in fluids, such as advection of pebbles in a river. In the case of heat transfer in fluids, where transport by advection in a fluid is always also accompanied by transport via heat diffusion (also known as heat conduction) the process of heat convection is understood to refer to the sum of heat transport by advection and diffusion/conduction.

Free, or natural, convection occurs when bulk fluid motions (streams and currents) are caused by buoyancy forces that result from density variations due to variations of temperature in the fluid. *Forced* convection is a term used when the streams and currents in the fluid are induced by external means—such as fans, stirrers, and pumps—creating an artificially induced convection current.

Convection-Cooling

Convective cooling is sometimes described as Newton's law of cooling:

The rate of heat loss of a body is proportional to the temperature difference between the body and its surroundings.

However, by definition, the validity of Newton's law of cooling requires that the rate of heat loss from convection be a linear function of ("proportional to") the temperature difference that drives heat transfer, and in convective cooling this is sometimes not the case. In general, convection is not linearly dependent on temperature gradients, and in some cases is strongly nonlinear. In these cases, Newton's law does not apply.

Convection vs. Conduction

In a body of fluid that is heated from underneath its container, conduction and convection can be considered to compete for dominance. If heat conduction is too great, fluid moving down by convection is heated by conduction so fast that its downward movement will be stopped due to its buoyancy, while fluid moving up by convection is cooled by conduction so fast that its driving buoyancy will diminish. On the other hand, if heat conduction is very low, a large temperature gradient may be formed and convection might be very strong.

The Rayleigh number (Ra) is a measure determining the relative strength of conduction and convection.

$$Ra = \frac{g\Delta\rho L^3}{\mu\alpha} = \frac{g\beta\Delta T L^3}{\nu\alpha}$$

where

- g is acceleration due to gravity,
- ρ is the density with $\Delta\rho$ being the density difference between the lower and upper ends,
- μ is the dynamic viscosity,
- α is the Thermal diffusivity,
- β is the volume thermal expansivity (sometimes denoted elsewhere),
- T is the temperature,
- ν is the kinematic viscosity, and
- L is characteristic length.

The Rayleigh number can be understood as the ratio between the rate of heat transfer by convection to the rate of heat transfer by conduction; or, equivalently, the ratio between the corresponding timescales (i.e. conduction timescale divided by convection timescale), up to a numerical factor. This can be seen as follows, where all calculations are up to numerical factors depending on the geometry of the system.

The buoyancy force driving the convection is roughly $g\Delta\rho L^3$, so the corresponding pressure is roughly $g\Delta\rho L$. In steady state, this is canceled by the shear stress due to viscosity, and therefore roughly equals $\mu V/L = \mu/T_{conv}$, where V is the typical fluid velocity due to convection and T_{conv} the order of its timescale. The conduction timescale, on the other hand, is of the order of $T_{cond} = L^2/\alpha$.

Convection occurs when the Rayleigh number is above 1,000–2,000.

Radiation

Thermal radiation occurs through a vacuum or any transparent medium (solid or fluid). It is the transfer of energy by means of photons in electromagnetic waves governed by the same laws. Earth's radiation balance depends on the incoming and the outgoing thermal radiation, Earth's energy budget. Anthropogenic perturbations in the climate system are responsible for a positive radiative forcing which reduces the net longwave radiation loss to space.

Thermal radiation is energy emitted by matter as electromagnetic waves, due to the pool of thermal energy in all matter with a temperature above absolute zero. Thermal radiation propagates without the presence of matter through the vacuum of space.

Red-hot iron object, transferring heat to the surrounding environment primarily through thermal radiation

Thermal radiation is a direct result of the random movements of atoms and molecules in matter. Since these atoms and molecules are composed of charged particles (protons and electrons), their movement results in the emission of electromagnetic radiation, which carries energy away from the surface.

The Stefan-Boltzmann equation, which describes the rate of transfer of radiant energy, is as follows for an object in a vacuum :

$$Q = \epsilon\sigma T^4$$

For radiative transfer between two objects, the equation is as follows:

$$Q = \epsilon\sigma(T_a^4 - T_b^4)$$

where Q is the heat flux, ε is the emissivity (unity for a black body), σ is the Stefan-Boltzmann constant, and T is the absolute temperature (in Kelvin or Rankine). Radiation is typically only important for very hot objects, or for objects with a large temperature difference.

Radiation from the sun, or solar radiation, can be harvested for heat and power. Unlike conductive and convective forms of heat transfer, thermal radiation can be concentrated in a small spot by using reflecting mirrors, which is exploited in concentrating solar power generation. For example, the sunlight reflected from mirrors heats the PS10 solar power tower and during the day it can heat water to 285 °C (545 °F).

Phase Transition

Phase transition or phase change, takes place in a thermodynamic system from one phase or state of matter to another one by heat transfer. Phase change examples are the melting of ice or the boiling of water. The Mason equation explains the growth of a water droplet based on the effects of heat transport on evaporation and condensation.

Lightning is a highly visible form of energy transfer and is an example of plasma present at Earth's surface. Typically, lightning discharges 30,000 amperes at up to 100 million volts, and emits light, radio waves, X-rays and even gamma rays. Plasma temperatures in lightning can approach 28,000 Kelvin (27,726.85 °C) (49,940.33 °F) and electron densities may exceed 10^{24} m^{-3}.

Types of phase transition occurring in the four fundamental states of matter, include:

- Solid - Deposition, freezing and solid to solid transformation.
- Gas - Boiling/evaporation, recombination/deionization, and sublimation.
- Liquid - Condensation and melting/fusion.
- Plasma - Ionization.

Boiling

The boiling point of a substance is the temperature at which the vapor pressure of the liquid equals the pressure surrounding the liquid and the liquid evaporates resulting in an abrupt change in vapor volume.

Saturation temperature means boiling point. The saturation temperature is the temperature for a corresponding saturation pressure at which a liquid boils into its vapor phase. The liquid can be said to be saturated with thermal energy. Any addition of thermal energy results in a phase transition.

Nucleate boiling of water.

At standard atmospheric pressure and low temperatures, no boiling occurs and the heat transfer rate is controlled by the usual single-phase mechanisms. As the surface temperature is increased, local boiling occurs and vapor bubbles nucleate, grow into the surrounding cooler fluid, and collapse. This is *sub-cooled nucleate boiling*, and is a very efficient heat transfer mechanism. At high bubble generation rates, the bubbles begin to interfere and the heat flux no longer increases rapidly with surface temperature (this is the departure from nucleate boiling, or DNB).

At similar standard atmospheric pressure and high temperatures, the hydrodynamically-quieter regime of film boiling is reached. Heat fluxes across the stable vapor layers are low, but rise slowly with temperature. Any contact between fluid and the surface that may be seen probably leads to the extremely rapid nucleation of a fresh vapor layer ("spontaneous nucleation"). At higher temperatures still, a maximum in the heat flux is reached (the critical heat flux, or CHF).

The Leidenfrost Effect demonstrates how nucleate boiling slows heat transfer due to gas bubbles on the heater's surface. As mentioned, gas-phase thermal conductivity is much lower than liquid-phase thermal conductivity, so the outcome is a kind of "gas thermal barrier".

Condensation

Condensation occurs when a vapor is cooled and changes its phase to a liquid. During condensation, the latent heat of vaporization must be released. The amount of the heat is the same as that absorbed during vaporization at the same fluid pressure.

There are several types of condensation:

- Homogeneous condensation, as during a formation of fog.
- Condensation in direct contact with subcooled liquid.
- Condensation on direct contact with a cooling wall of a heat exchanger: This is the most common mode used in industry:
    - Filmwise condensation is when a liquid film is formed on the subcooled surface, and usually occurs when the liquid wets the surface.
    - Dropwise condensation is when liquid drops are formed on the subcooled surface, and usually occurs when the liquid does not wet the surface.

Dropwise condensation is difficult to sustain reliably; therefore, industrial equipment is normally designed to operate in filmwise condensation mode.

Melting

Melting is a physical process that results in the phase transition of a substance from a solid to a liquid. The internal energy of a substance is increased, typically by the application of heat or pressure, resulting in a rise of its temperature to the melting point, at which the ordering of ionic or molecular entities in the solid breaks down to a less ordered state and the solid liquefies. An object that has melted completely is molten.

Ice melting

Substances in the molten state generally have reduced viscosity with elevated temperature; an exception to this maxim is the element sulfur, whose viscosity increases to a point due to polymerization and then decreases with higher temperatures in its molten state.

Modeling Approaches

Heat transfer can be modeled in the following ways.

Climate Models

Climate models study the radiant heat transfer by using quantitative methods to simulate the interactions of the atmosphere, oceans, land surface, and ice.

Heat Equation

The heat equation is an important partial differential equation that describes the distribution of heat (or variation in temperature) in a given region over time. In some cases, exact solutions of the equation are available; in other cases the equation must be solved numerically using computational methods.

Lumped System Analysis

Lumped system analysis often reduces the complexity of the equations to one first-order linear differential equation, in which case heating and cooling are described by a simple exponential solution, often referred to as Newton's law of cooling.

System analysis by the lumped capacitance model is a common approximation in transient conduction that may be used whenever heat conduction within an object is much faster than heat conduction across the boundary of the object. This is a method of approximation that reduces one aspect of the transient conduction system—that within the object—to an equivalent steady state system. That is, the method assumes that the temperature within the object is completely uniform, although its value may be changing in time.

In this method, the ratio of the conductive heat resistance within the object to the convective heat transfer resistance across the object's boundary, known as the *Biot number*, is calculated. For small Biot numbers, the approximation of *spatially uniform temperature within the object* can be used: it can be presumed that heat transferred into the object has time to uniformly distribute itself, due to the lower resistance to doing so, as compared with the resistance to heat entering the object.

Engineering

Heat transfer has broad application to the functioning of numerous devices and systems. Heat-transfer principles may be used to preserve, increase, or decrease temperature in a wide variety of circumstances.

Heat exposure as part of a fire test for firestop products

Heat transfer methods are used in numerous disciplines, such as automotive engineering, thermal management of electronic devices and systems, climate control, insulation, materials processing, and power station engineering.

Insulation, Radiance and Resistance

Thermal insulators are materials specifically designed to reduce the flow of heat by limiting conduction, convection, or both. Thermal resistance is a heat property and the measurement by which an object or material resists to heat flow (heat per time unit or thermal resistance) to temperature difference.

Radiance or spectral radiance are measures of the quantity of radiation that passes through or is emitted. Radiant barriers are materials that reflect radiation, and therefore reduce the flow of heat from radiation sources. Good insulators are not necessarily good radiant barriers, and vice versa. Metal, for instance, is an excellent reflector and a poor insulator.

The effectiveness of a radiant barrier is indicated by its reflectivity, which is the fraction of radiation reflected. A material with a high reflectivity (at a given wavelength) has a low emissivity (at that same wavelength), and vice versa. At any specific wavelength, reflectivity = 1 - emissivity. An ideal radiant barrier would have a reflectivity of 1, and would therefore reflect 100 percent of incoming radiation. Vacuum flasks, or Dewars, are silvered to approach this ideal. In the vacuum of space, satellites use multi-layer insulation, which consists of many layers of aluminized (shiny) Mylar to greatly reduce radiation heat transfer and control satellite temperature.

Devices

- Heat engine is a system that performs the conversion of heat or thermal energy to mechanical energy which can then be used to do mechanical work.

- Thermocouple is a temperature-measuring device and widely used type of temperature sensor for measurement and control, and can also be used to convert heat into electric power.
- Thermoelectric cooler is a solid state electronic device that pumps (transfers) heat from one side of the device to the other when electric current is passed through it. It is based on the Peltier effect.
- Thermal diode or thermal rectifier is a device that causes heat to flow preferentially in one direction.

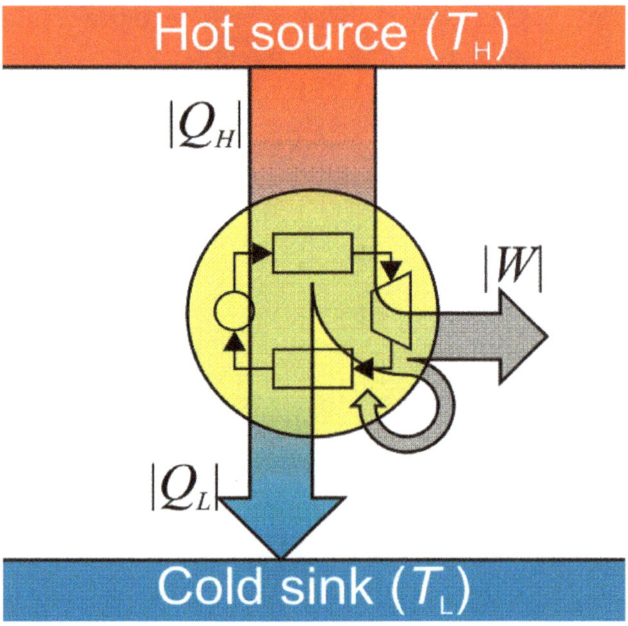

Heat engine diagram

Heat Exchangers

A heat exchanger is used for more efficient heat transfer or to dissipate heat. Heat exchangers are widely used in refrigeration, air conditioning, space heating, power generation, and chemical processing. One common example of a heat exchanger is a car's radiator, in which the hot coolant fluid is cooled by the flow of air over the radiator's surface.

Common types of heat exchanger flows include parallel flow, counter flow, and cross flow. In parallel flow, both fluids move in the same direction while transferring heat; in counter flow, the fluids move in opposite directions; and in cross flow, the fluids move at right angles to each other. Common constructions for heat exchanger include shell and tube, double pipe, extruded finned pipe, spiral fin pipe, u-tube, and stacked plate.

A heat sink is a component that transfers heat generated within a solid material to a fluid medium, such as air or a liquid. Examples of heat sinks are the heat exchangers used in refrigeration and air conditioning systems or the radiator in a car. A heat pipe is another heat-transfer device that combines thermal conductivity and phase transition to efficiently transfer heat between two solid interfaces.

Examples

Architecture

Efficient energy use is the goal to reduce the amount of energy required in heating or cooling. In architecture, condensation and air currents can cause cosmetic or structural damage. An energy audit can help to assess the implementation of recommended corrective procedures. For instance, insulation improvements, air sealing of structural leaks or the addition of energy-efficient windows and doors.

- Smart meter is a device that records electric energy consumption in intervals.
- Thermal transmittance is the rate of transfer of heat through a structure divided by the difference in temperature across the structure. It is expressed in watts per square meter per kelvin, or W/m^2K. Well-insulated parts of a building have a low thermal transmittance, whereas poorly-insulated parts of a building have a high thermal transmittance.
- Thermostat is a device to monitor and control temperature.

Climate Engineering

Climate engineering consists of carbon dioxide removal and solar radiation management. Since the amount of carbon dioxide determines the radiative balance of Earth atmosphere, carbon dioxide removal techniques can be applied to reduce the radiative forcing. Solar radiation management is the attempt to absorb less solar radiation to offset the effects of greenhouse gases.

An example application in climate engineering includes the creation of Biochar through the pyrolysis process. Thus, storing greenhouse gases in carbon reduces the radiative forcing capacity in the atmosphere, causing more long-wave (infrared) radiation out to Space.

Greenhouse Effect

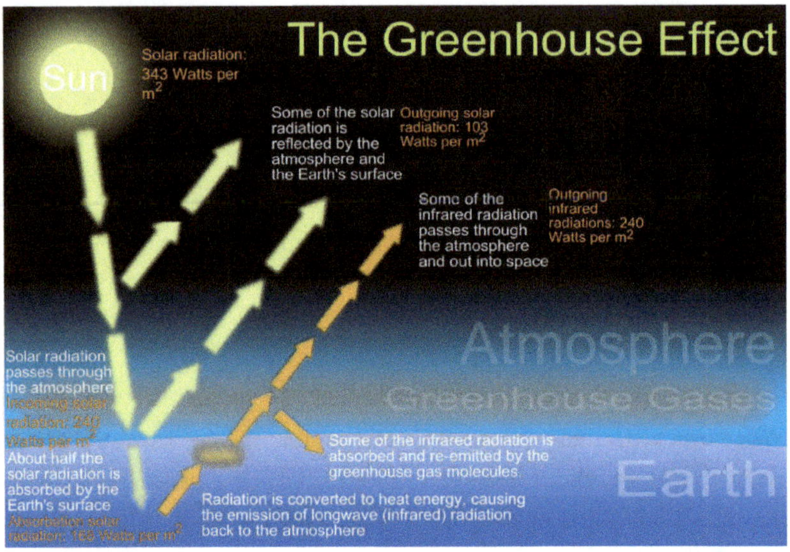

A representation of the exchanges of energy between the source (the Sun), the Earth's surface, the Earth's atmosphere, and the ultimate sink outer space. The ability of the atmosphere to capture and recycle energy emitted by the Earth surface is the defining characteristic of the greenhouse effect.

The greenhouse effect is a process by which thermal radiation from a planetary surface is absorbed by atmospheric greenhouse gases, and is re-radiated in all directions. Since part of this re-radiation is back towards the surface and the lower atmosphere, it results in an elevation of the average surface temperature above what it would be in the absence of the gases.

Heat Transfer in the Human Body

The principles of heat transfer in engineering systems can be applied to the human body in order to determine how the body transfers heat. Heat is produced in the body by the continuous metabolism of nutrients which provides energy for the systems of the body. The human body must maintain a consistent internal temperature in order to maintain healthy bodily functions. Therefore, excess heat must be dissipated from the body to keep it from overheating. When a person engages in elevated levels of physical activity, the body requires additional fuel which increases the metabolic rate and the rate of heat production. The body must then use additional methods to remove the additional heat produced in order to keep the internal temperature at a healthy level.

Heat transfer by convection is driven by the movement of fluids over the surface of the body. This convective fluid can be either a liquid or a gas. For heat transfer from the outer surface of the body, the convection mechanism is dependent on the surface area of the body, the velocity of the air, and the temperature gradient between the surface of the skin and the ambient air. The normal temperature of the body is approximately 37 °C. Heat transfer occurs more readily when the temperature of the surroundings is significantly less than the normal body temperature. This concept explains why a person feels “cold” when not enough covering is worn when exposed to a cold environment. Clothing can be considered an insulator which provides thermal resistance to heat flow over the covered portion of the body. This thermal resistance causes the temperature on the surface of the clothing to be less than the temperature on the surface of the skin. This smaller temperature gradient between the surface temperature and the ambient temperature will cause a lower rate of heat transfer than if the skin were not covered.

In order to ensure that one portion of the body is not significantly hotter than another portion, heat must be distributed evenly through the bodily tissues. Blood flowing through blood vessels acts as a convective fluid and helps to prevent any buildup of excess heat inside the tissues of the body. This flow of blood through the vessels can be modeled as pipe flow in an engineering system. The heat carried by the blood is determined by the temperature of the surrounding tissue, the diameter of the blood vessel, the thickness of the fluid, velocity of the flow, and the heat transfer coefficient of the blood. The velocity, blood vessel diameter, and the fluid thickness can all be related with the Reynolds Number, a dimensionless number used in fluid mechanics to characterize the flow of fluids.

Latent heat loss, also known as evaporative heat loss, accounts for a large fraction of heat loss from the body. When the core temperature of the body increases, the body triggers sweat glands in the skin to bring additional moisture to the surface of the skin. The liquid is then transformed into vapor which removes heat from the surface of the body. The rate of evaporation heat loss is directly related to the vapor pressure at the skin surface and the amount of moisture present on the skin. Therefore, the maximum of heat transfer will occur when the skin is completely wet. The body continuously loses water by evaporation but the most significant amount of heat loss occurs during periods of increased physical activity.

Cooling Techniques

Evaporative Cooling

Evaporative cooling happens when water vapor is added to the surrounding air. The energy needed to evaporate the water is taken from the air in the form of sensible heat and converted into latent heat, while the air remains at a constant enthalpy. Latent heat describes the amount of heat that is needed to evaporate the liquid; this heat comes from the liquid itself and the surrounding gas and surfaces.

A traditional air cooler in Mirzapur, Uttar Pradesh, India

The greater the difference between the two temperatures, the greater the evaporative cooling effect. When the temperatures are the same, no net evaporation of water in air occurs; thus, there is no cooling effect.

Laser Cooling

In Quantum Physics laser cooling is used to achieve temperatures of near absolute zero (−273.15 °C, −459.67 °F) of atomic and molecular samples, to observe unique quantum effects that can only occur at this heat level.

- Doppler cooling is the most common method of laser cooling.
- Sympathetic cooling is a process in which particles of one type cool particles of another type. Typically, atomic ions that can be directly laser-cooled are used to cool nearby ions or atoms. This technique allows cooling of ions and atoms that cannot be laser cooled directly.

Magnetic Cooling

Magnetic evaporative cooling is a process for lowering the temperature of a group of atoms, after pre-cooled by methods such as laser cooling. Magnetic refrigeration cools below 0.3K, by making use of the magnetocaloric effect.

Radiative Cooling

Radiative cooling is the process by which a body loses heat by radiation. Outgoing energy is an important effect in the Earth's energy budget. In the case of the Earth-atmosphere system, it refers to the process by which long-wave (infrared) radiation is emitted to balance the absorption of short-wave (visible) energy from the Sun. Convective transport of heat and evaporative transport of latent heat both remove heat from the surface and redistribute it in the atmosphere.

Thermal Energy Storage

Thermal energy storage refers to technologies used to collect and store energy for later use. They can be employed to balance energy demand between day and nighttime. The thermal reservoir may be maintained at a temperature above (hotter) or below (colder) than that of the ambient environment. Applications include later use in space heating, domestic or process hot water, or to generate electricity.

Heat Transfer Enhancement

Heat transfer enhancement is the process of increasing the effectiveness of heat exchangers. This can be achieved when the heat transfer power of a given device is increased or when the pressure losses generated by the device are reduced. A variety of techniques can be applied to this effect, including generating strong secondary flows or increasing boundary layer turbulence.

Principle

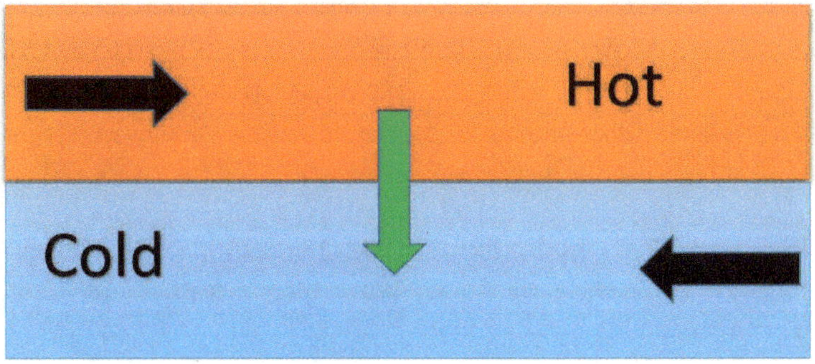

Two fluid counter-flow

During the earliest attempts to enhance heat transfer, plain (or smooth) surfaced were used. This surface requires a special surface geometry able to provide higher hA values per unit surface area in comparison with a plain surface. The ratio of hA of an enhanced heat transfer surface to the plain surface is called Enhancement Ratio " E_h". Thus,

$$E_h = \frac{hA}{(hA)_p}$$

The heat transfer rate for a two-fluid counterflow heat exchanger is given by

$$Q = UA\Delta T_m$$

In order to better illustrate the benefits of enhancement, the total length 'L' of the tube is multiplied and divided in the equation

$$Q = \frac{UA}{L} L\Delta T_m$$

Where $\frac{L}{UA}$ is the overall thermal resistance per unit tube length. And it is given by

$$\frac{L}{UA} = \frac{L}{\eta_1 h_1 A_1} + \frac{Lt_w}{k_w A_m} + \frac{L}{\eta_2 h_2 A_2}$$

The subscripts 1 and 2, describe the two different fluids. The surface efficiency is represented by η employing extended surfaces. One aspect to take into consideration is that the latter equation does not include any fouling resistances due to its simplicity, which can be important. In order to enhance the performance of the heat exchanger, the term, $\frac{UA}{L}$ must be increased. For achieving a reduced thermal resistance, the enhanced surface geometry may be used to increase one or both terms $\frac{hA}{L}$ in relation to the plain surfaces, leading to a reduced thermal resistance per unit tube length, $\frac{L}{UA}$. This reduced term may be used to achieve one of the following three objectives:

1. Size reduction. Maintaining the heat exchange rate Q constant, the length of the heat exchanger may be reduced, providing a heat exchanger of smaller proportions.
2. Increased UA.
    - Reduced Δt_m: maintaining both Q and the length constant, Δt_m: can be reduced increasing thermodynamic process efficiency leading to reduced operation costs.
    - Increased heat exchange: Increasing $\frac{UA}{L}$ and keeping a constant length will lead to an increased Q for fixed fluid inlet temperature.
3. Reduced pumping power for fixed heat duty. This will require smaller velocities of operation than the plain surface and a normally not desired, increased frontal area.

Depending on the objectives for the design, any of the three different performance improvements can be used on an enhanced surface.

Internal Flow

There are several available options for enhancing heat transfer. The enhancement can be achieved by increasing the surface area for convection or/and increasing the convection coefficient. For example, the surface roughness can be used to increase h in order to enhance turbulence. This can be achieved through machining or other kinds of insertions like coil-spring wire. The insert provides a helical roughness in contact with the surface.

Coil spring wire insert

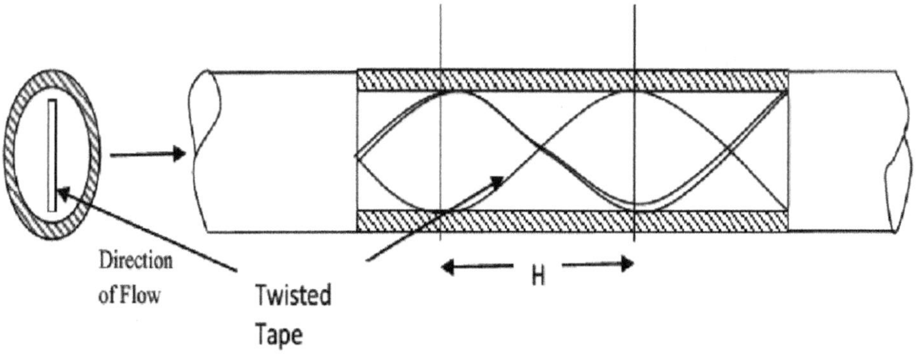

Twisted tape Tube Insert

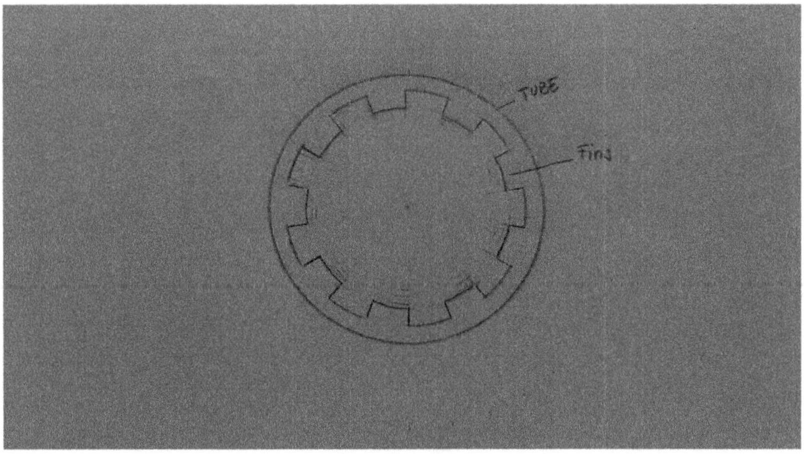

Longitudinal fins

Helical ribs

The convection coefficient may also be increased by an insert of a twisted tape that consists in a periodical twist through 360 degrees. Tangential inserts optimize the velocity of the flow near the tube wall, while providing a bigger heat transfer area. While, increased area and convection coefficient can be achieved by applying spiral fin or ribs inserts. Other aspects such pressure drop must be taken into consideration in order to meet the fan or pump power constraints.

Helically Coiled Tube

The coil spring insert may enhance heat transfer without turbulence or additional heat transfer surface area. A secondary flow is induces the fluid creating two longitudinal vortices. This could result, (in contrast to a right tube) in highly non-uniform local h around the periphery of the tube.

Leading to a dependence of the local heat transfer coefficients on the different locations along the tube (x). Supposing that the conditions for the heat flux are constant, the mean fluid temperature, $T_m(x)$ can be estimated as follows,

$$T_m(x) = T_{m,i} + \frac{q_s'' P}{\dot{m} c_p} x \quad \text{where} \quad q_s'' = \text{constant.}$$

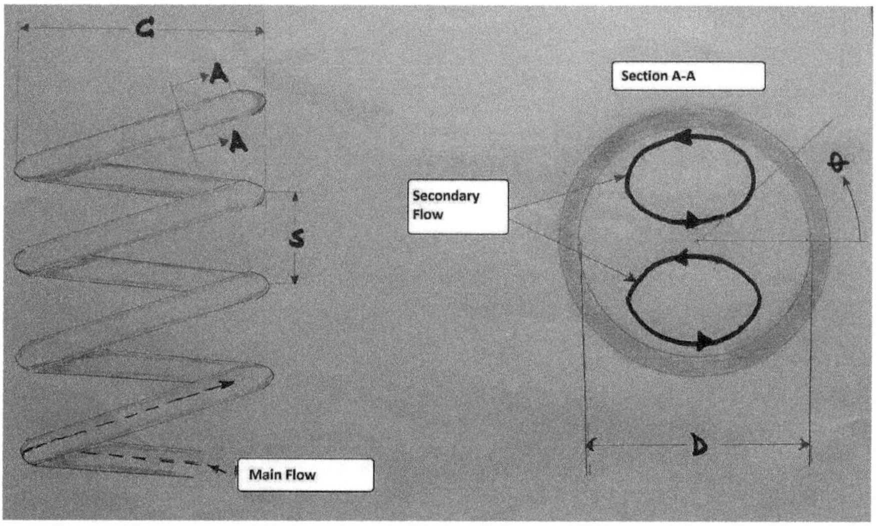

Helically coiled tube with secondary flow

Maximum fluid temperatures near the tube wall are present when the fluid is heated, and because the heat transfer coefficient is strongly depended of angle (ϕ), the calculation of the maximum local temperature is not straight forward. For this purpose, correlations for the peripherally averaged Nusselt number are, if none, of little use when keeping heat flux conditions constant. On the other hand, correlations for the peripherally averaged Nusselt number for constant wall temperature are very useful.

The secondary flow:

- Increases heat transfer rates.
- Increases friction loses.
- Decreases entrance length.
- Reduces the difference between the laminar and turbulent heat transfer rates, in contrast to the straight tube case.

The coil pitch S has negligible influence on the pressure drop and the heat transfer rates. For the helical tube, the critical Reynolds number to the onset of turbulence is,

$$Re_{D,c,h} = Re_{D,c}[1 + 12(D/C)^{0.5}]$$

where $Re_{D,c}$ is given by $Re_{D,c} \approx 2300$ in turbulent and fully developed state.

The delays on the transition from laminar to turbulent state are strongly dependent on strong secondary flows associated with tightly wound helically coiled tubes. The friction factor for fully developed laminar flow with $C/D \geq 3$ is,

$$f = \frac{64}{Re_D} \quad \text{where} \quad Re_D(D/C)^{0.5} \leq 30.$$ C is the outer diameter of the helical coil.

and

$$f = \frac{27}{Re_D^{0.725}}(D/C)^{0.1375} \quad \text{for} \quad 30 \leq Re_D(D/C)^{0.5} \leq 300$$

and

$$f = \frac{7.2}{Re_D^{0.5}}(D/C)^{0.25} \quad \text{where} \quad 300 \leq Re_D(D/C)^{0.5}$$

For cases where $C/D \leq 3$, there is available recommendations provided by Shah and Joshi. The heat transfer coefficient may be used in the equation for the Newton's law of cooling equation

$$q''_s = h(T_s - T_m)$$

and can be evaluated from the correlation,

$$Nu_D = \left[\left(3.66 + \frac{4.343}{a}\right)^3 + 1.158\left(\frac{Re_D(D/C)^{0.5}}{b}\right)^{\frac{3}{2}}\right]^{\frac{1}{3}} \left(\frac{\mu}{\mu_s}\right)^{0.14}$$

$$\text{where} \quad a = \left(1 + \frac{927(C/D)}{Re_D^2 Pr}\right) \quad \text{and} \quad b = 1 + \frac{0.477}{Pr}$$

The correlations for the friction factor in turbulent state are based in limited data. Increased heat transfer due to the secondary flow is not significant in turbulent state constituting less than 10% for $C/D \geq 20$. Furthermore, augmentation created by the use of helically coiled tubes due to the secondary flow is usually employed only for situations where the flow is in the laminar state. In this state, the entrance length is 20% to 50% shorter in comparison with the straight tube. In the case of turbulent flow, the flow becomes fully developed during the first half-turn of the helically coiled tube. For this reason, the entrance region can be neglected in many engineering calculations. If the liquid or gas is heated in a straight tube, the fluid that passes near the centerline, will exit the tube in a much shorter time and will always be cooler than the fluid passing near the wall.

MASS TRANSFER

Mass transfer is the net movement of mass from one location, usually meaning stream, phase, fraction or component, to another. Mass transfer occurs in many processes, such as absorption, evaporation, drying, precipitation, membrane filtration, and distillation. Mass transfer is used by different scientific disciplines for different processes and mechanisms. The phrase is commonly used in engineering for physical processes that involve diffusive and convective transport of chemical species within physical systems.

Some common examples of mass transfer processes are the evaporation of water from a pond to the atmosphere, the purification of blood in the kidneys and liver, and the distillation of alcohol. In industrial processes, mass transfer operations include separation of chemical components in distillation columns, absorbers such as scrubbers or stripping, adsorbers such as activated carbon beds, and liquid-liquid extraction. Mass transfer is often coupled to additional transport processes, for instance in industrial cooling towers. These towers couple heat transfer to mass transfer by allowing hot water to flow in contact with hotter air and evaporate as it absorbs heat from the air.

Astrophysics

In astrophysics, mass transfer is the process by which matter gravitationally bound to a body, usually a star, fills its Roche lobe and becomes gravitationally bound to a second body, usually a compact object (white dwarf, neutron star or black hole), and is eventually accreted onto it. It is a common phenomenon in binary systems, and may play an important role in some types of supernovae and pulsars.

Chemical Engineering

Mass transfer finds extensive application in chemical engineering problems. It is used in reaction engineering, separations engineering, heat transfer engineering, and many other sub-disciplines of chemical engineering.

The driving force for mass transfer is typically a difference in chemical potential, when it can be defined, though other thermodynamic gradients may couple to the flow of mass and drive it as well. A chemical species moves from areas of high chemical potential to areas of low chemical potential. Thus, the maximum theoretical extent of a given mass transfer is typically determined by the point at which the chemical potential is uniform. For single phase-systems, this usually translates to uniform concentration throughout the phase, while for multiphase systems chemical species will often prefer one phase over the others and reach a uniform chemical potential only when most of the chemical species has been absorbed into the preferred phase, as in liquid-liquid extraction.

While thermodynamic equilibrium determines the theoretical extent of a given mass transfer operation, the actual rate of mass transfer will depend on additional factors including the flow patterns within the system and the diffusivities of the species in each phase. This rate can be quantified through the calculation and application of mass transfer coefficients for an overall process. These mass transfer coefficients are typically published in terms of dimensionless numbers, often including Péclet numbers, Reynolds numbers, Sherwood numbers and Schmidt numbers, among others.

Analogies Between Heat, Mass, and Momentum Transfer

There are notable similarities in the commonly used approximate differential equations for momentum, heat, and mass transfer. The molecular transfer equations of Newton's law for fluid momentum at low Reynolds number (Stokes flow), Fourier's law for heat, and Fick's law for mass are very similar, since they are all linear approximations to transport of conserved quantities in a flow field. At higher Reynolds number, the analogy between mass and heat transfer and momentum transfer becomes less useful due to the nonlinearity of the Navier-Stokes equation (or more fundamentally, the general momentum conservation equation), but the analogy between heat and mass transfer remains good. A great deal of effort has been devoted to developing analogies among these three transport processes so as to allow prediction of one from any of the others.

REFERENCES

1. "Arther Dehon Little". Scatter Acorns That Oaks May Grow. MIT Libraries. Retrieved 13 November 2013.
2. "Arthur D. Little, William H. Walker, and Warren K. Lewis". Chemical Heritage Foundation. Retrieved 13 November 2013.
3. Zaghloul, Mofreh R.; Bourham, Mohamed A.; Doster, J.Michael (April 2000). "Energy-averaged electron–ion momentum transport cross section in the Born approximation and Debye–Hückel potential: Comparison with the cut-off theory". Physics Letters A. 268 (4-6): 375–381.
4. Bransden, B.H.; Joachain, C.J. (2003). Physics of atoms and molecules (2. ed.). Harlow [u.a.]: Prentice-Hall. p. 584. ISBN 978-0582356924.
5. Paul A., Tipler; Gene Mosca (2008). Physics for Scientists and Engineers, Volume 1 (6th ed.). New York, NY: Worth Publishers. pp. 666–670. ISBN 1-4292-0132-0.
6. New Jersey Institute of Technology, Chemical Engineering Dept. "B.S. Chemical Engineering". NJIT. Retrieved 9 April 2011.
7. Lienhard, John H.,V; Lienhard, John H., V (2008). A Heat Transfer Textbook (3rd ed.). Cambridge, Massachusetts: Phlogiston Press. ISBN 978-0-9713835-3-1.
8. Welty, James R.; Wicks, Charles E.; Wilson, Robert Elliott (1976). Fundamentals of momentum, heat, and mass transfer (2 ed.). New York: Wiley. ISBN 978-0-471-93354-0. OCLC 2213384.
9. Faghri, Amir; Zhang, Yuwen; Howell, John (2010). Advanced Heat and Mass Transfer. Columbia, MO: Global Digital Press. ISBN 978-0-9842760-0-4.
10. Taylor, R.A. (2012). "Socioeconomic impacts of heat transfer research". International Communications in Heat and Mass Transfer. 39 (10): 1467–1473.
11. "Thermal-FluidsPedia | Mass transfer"
12. Abbott, J.M. Smith, H.C. Van Ness, M.M. (2005). Introduction to chemical engineering thermodynamics (7th ed.). Boston ; Montreal: McGraw-Hill. ISBN 0-07-310445-0.
13. "Thermal-FluidsPedia | Heat conduction".
14. Çengel, Yunus (2003). Heat Transfer: a practical approach. McGraw-Hill series in mechanical engineering. (2nd ed.). Boston: McGraw-Hill. ISBN 978-0-07-245893-0. OCLC 300472921. Retrieved 2009-04-20.
15. «Thermal-FluidsPedia | Convective heat transfer»

16. «Convection — Heat Transfer». Engineers Edge. Engineers Edge. Retrieved 2009-04-20.
17. Geankoplis, Christie John (2003). Transport processes and separation process principles : (includes unit operations) (4th ed.). Upper Saddle River, NJ: Prentice Hall Professional Technical Reference. ISBN 0-13-101367-X.
18. «Thermal-FluidsPedia | Radiation»
19. Mojiri, A (2013). «Spectral beam splitting for efficient conversion of solar energy—A review». Renewable and Sustainable Energy Reviews. 28: 654–663. doi:10.1016/j.rser.2013.08.026.
20. Taylor, R.A., Applicability of Nanofluids in High Flux Solar Collectors JOURNAL OF RENEWABLE AND SUSTAINABLE ENERGY 3, 023104, 2011
21. See Flashes in the Sky: Earth›s Gamma-Ray Bursts Triggered by Lightning
22. David.E. Goldberg (1988). 3,000 Solved Problems in Chemistry (1st ed.). McGraw-Hill. ISBN 0-07-023684-4. Section 17.43, page 321
23. Louis Theodore, R. Ryan Dupont and Kumar Ganesan (Editors) (1999). Pollution Prevention: The Waste Management Approach to the 21st Century. CRC Press. ISBN 1-56670-495-2. Section 27, page 15
24. C.Michael Hogan (2011) Sulfur, Encyclopedia of Earth, eds. A.Jorgensen and C.J.Cleveland, National Council for Science and the environment, Washington DC
25. Wendl, MC (2012). Theoretical Foundations of Conduction and Convection Heat Transfer. Wendl Foundation.
26. «How to simplify for small Biot numbers». Retrieved December 21, 2016.
27. Fundamentals of Classical Thermodynamics, 3rd ed. p. 159, (1985) by G. J. Van Wylen and R. E. Sonntag.
28. Mechanical efficiency of heat engines, p. 1 (2007) by James R. Senf: «Heat engines are made to provide mechanical energy from thermal energy.»
29. «EnergySavers: Tips on Saving Money & Energy at Home» (PDF). U.S. Department of Energy. Retrieved March 2, 2012.
30. Hartman,Carl and Bibb, Lewis. «The Human body and its enemies: a textbook of physiology hygiene and sanitation», World Book Co., 1913, p.232.
31. Cengel, Yunus A. and Ghajar, Afshin J. «Heat and Mass Transfer: Fundamentals and Applications.» , McGraw-Hill, 4th Edition, 2010.
32. Tao, Xiaoming. «Smart fibres, fabrics, and clothing» , Woodhead Publishing, 2001
33. Wilmore, Jack H.; Costill, David L.; Kenney, Larry. «Physiology of sport and exercise». Human Kinetics. 2008: 256.
34. Webb, Kim, Ralph L., Nae-Hyun (June 23, 2005). Principles of Enhanced Heat Transfer. CRC Press; 2 edition. ISBN 978-1591690146.
35. Shah, R. K., and S.D. Joshi, in Handbook of Single-phase Convective Heat transfer, Chap. 5, Wiley-Interscience, Hoboken, NJ, 1987

36. Incropera, Dewitt, Bergman, Lavine, Frank P., David P., Theodore L., Adrienne S. (2013). Principles of Heat and Mass Transfer,. John Wiley & Sons; 7th Edition, Interna edition. ISBN 978-0470646151.
37. Welty, James R.; Wicks, Charles E.; Wilson, Robert Elliott (1976). Fundamentals of momentum, heat, and mass transfer (2 ed.). Wiley.
38. Bird, R.B.; Stewart, W.E.; Lightfoot, E.N. (2007). Transport Phenomena (2 ed.). Wiley.
39. Taylor, R.; Krishna, R. (1993). Multicomponent Mass Transfer. Wiley.

CHAPTER 4

Unit Processes

Chapter Contents

In unit process there is no any physical changes but chemical changes occur. Unit process includes all the treatment methods in which the removal of contaminants is brought about by chemical or biological reactions. It is considered as a secondary treatment.

Unit process involves a chemical change or sometime it referred as chemical changes along with physical change example production of paracetamol from benzene.

INTRODUCTION

A ''unit process'' is one or more grouped operations in a manufacturing system that can be defined and separated from others.

In life-cycle assessment (LCA) and ISO 14040, a unit process is defined as "smallest element considered in the life cycle inventory analysis for which input and output data are quantified".

CHEMICAL REACTION ENGINEERING

Chemical reaction engineering (reaction engineering or reactor engineering) is a specialty in chemical engineering or industrial chemistry dealing with chemical reactors. Frequently the term relates specifically to catalytic reaction systems where either a homogeneous or heterogeneous catalyst is present in the reactor. Sometimes a reactor per se is not present by itself, but rather is integrated into a process, for example in reactive separations vessels, retorts, certain fuel cells, and photocatalytic surfaces. The issue of solvent effects on reaction kinetics is also considered as an integral part.

Origin of Chemical Reaction Engineering

Chemical reaction engineering as a discipline started in the early 1950s under the impulse of researchers at the Shell Amsterdam research center and the university of Delft. The term chemical reaction engineering was apparently coined by J.C. Vlugter while preparing the 1st European Symposium on Chemical Reaction Engineering which was held in Amsterdam in 1957.

Discipline

Chemical reaction engineering aims at studying and optimizing chemical reactions in order to define the best reactor design. Hence, the interactions of flow phenomena, mass transfer, heat transfer, and reaction kinetics are of prime importance in order to relate reactor performance to feed composition and operating conditions. Although originally applied to the petroleum and petrochemical industries, its general methodology combining reaction chemistry and chemical engineering concepts allows to optimize a variety of systems where modeling or engineering of reactions is needed. Chemical reaction engineering approaches are indeed tailored for the development of new processes and the improvement of existing technologies.

ISCRE Symposia

The most important series of symposia are the International Symposia on Chemical Reaction Engineering or ISCRE conferences. These three-day conferences are held every two years, rotating among sites in North America, Europe, and the Asia-Pacific region, on a six-year cycle. ISCRE-24 will be held in Minneapolis, MN, on June 12–15, 2016. These conferences bring together for three days distinguished international researchers in reaction engineering, prominent industrial practitioners, and new researchers and students of this multifaceted field. ISCRE symposia are a unique gathering place for reaction engineers where research gains are consolidated and new frontiers explored. The state of the art of various sub-disciplines of reaction engineering is reviewed in a timely manner, and new research initiatives are discussed.

CHEMICAL KINETICS

Chemical kinetics, also known as reaction kinetics, is the study of rates of chemical processes. Chemical kinetics includes investigations of how different experimental conditions can influence the speed of a chemical reaction and yield information about the reaction's mechanism and transition states, as well as the construction of mathematical models that can describe the characteristics of a chemical reaction.

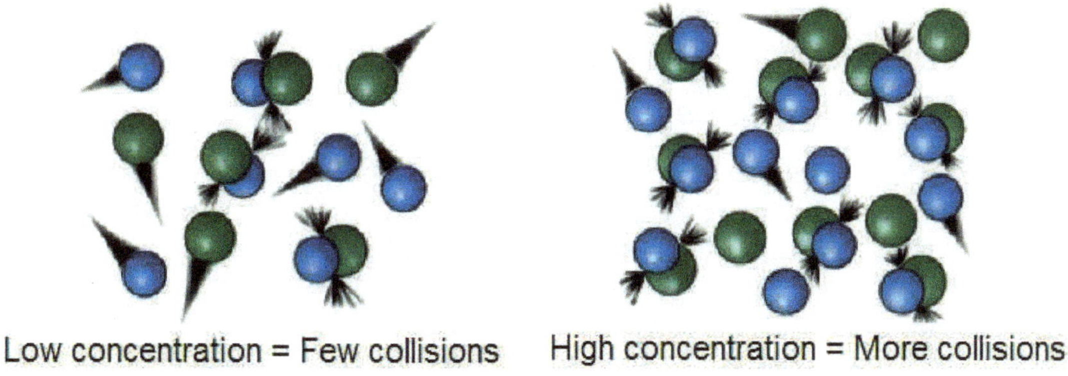

Reaction rate tends to increase with concentration – a phenomenon explained by collision theory.

History

In 1864, Peter Waage and Cato Guldberg pioneered the development of chemical kinetics by formulating the law of mass action, which states that the speed of a chemical reaction is proportional to the quantity of the reacting substances.

Van ‘t Hoff studied chemical dynamics and published in 1884 his famous “Etudes de dynamique chimique”. In 1901 he was awarded by the first Nobel Prize in Chemistry “in recognition of the extraordinary services he has rendered by the discovery of the laws of chemical dynamics and osmotic pressure in solutions”. After van ‘t Hoff, chemical kinetics deals with the experimental determination of reaction rates from which rate laws and rate constants are derived. Relatively simple rate laws exist for zero order reactions (for which reaction rates are independent of concentration), first order reactions, and second order reactions, and can be derived for others. Elementary reactions follow the law of mass action, but the rate law of stepwise reactions has to be derived by combining the rate laws of the various elementary steps, and can become rather complex. In consecutive reactions, the rate-determining step often determines the kinetics. In consecutive first order reactions, a steady state approximation can simplify the rate law. The activation energy for a reaction is experimentally determined through the Arrhenius equation and the Eyring equation. The main factors that influence the reaction rate include: the physical state of the reactants, the concentrations of the reactants, the temperature at which the reaction occurs, and whether or not any catalysts are present in the reaction.

Gorban and Yablonsky have suggested that the history of chemical dynamics can be divided into three eras. The first is the van ‘t Hoff wave searching for the general laws of chemical reactions and relating kinetics to thermodynamics. The second may be called the Semenov--Hinshelwood wave with emphasis on reaction mechanisms, especially for chain reactions. The third is associated with Aris and the detailed mathematical description of chemical reaction networks.

Factors Affecting Reaction Rate

Nature of the Reactants

Depending upon what substances are reacting, the reaction rate varies. Acid/base reactions, the formation of salts, and ion exchange are fast reactions. When covalent bond formation takes place between the molecules and when large molecules are formed, the reactions tend to be very slow. Nature and strength of bonds in reactant molecules greatly influence the rate of its transformation into products.

Physical State

The physical state (solid, liquid, or gas) of a reactant is also an important factor of the rate of change. When reactants are in the same phase, as in aqueous solution, thermal motion brings them into contact. However, when they are in different phases, the reaction is limited to the interface between the reactants. Reaction can occur only at their area of contact; in the case of a liquid and a gas, at the surface of the liquid. Vigorous shaking and stirring may be needed to bring the reaction to completion. This means that the more finely divided a solid or liquid reactant the greater its surface area per unit volume and the more contact it with the other reactant, thus the faster the reaction. To make an analogy, for example, when one starts a fire, one uses wood chips and small branches — one does not start with large logs right away. In addition, ionic substances reacts faster due to the electrostatic attraction that exist between reacting substances, while in organic substances, the stable configuration of the connectivity between carbon atoms also makes the majority of the portions of the organic molecule non-polar and makes organic molecule reacts in a very slow phase with each other.

Surface Area of Solids

In a solid, only those particles that are at the surface can be involved in a reaction. Crushing a solid into smaller parts means that more particles are present at the surface, and the frequency of collisions between these and reactant particles increases, and so reaction occurs more rapidly. For example, Sherbet (powder) is a mixture of very fine powder of malic acid (a weak organic acid) and sodium hydrogen carbonate. On contact with the saliva in the mouth, these chemicals quickly dissolve and react, releasing carbon dioxide and providing for the fizzy sensation. Also, fireworks manufacturers modify the surface area of solid reactants to control the rate at which the fuels in fireworks are oxidized, using this to create different effects. For example, finely divided aluminum confined in a shell explodes violently. If larger pieces of aluminum are used, the reaction is slower and sparks are seen as pieces of burning metal are ejected.

Concentration

The reactions are due to collisions of reactant species. The frequency with which the molecules or ions collide depends upon their concentrations. The more crowded the molecules are, the more likely they are to collide and react with one another. Thus, an increase in the concentrations of the reactants will usually result in the corresponding increase in the reaction rate, while a decrease in the concentrations will usually have a reverse effect. For example, combustion will occur more rapidly in pure oxygen than in air (21% oxygen).

The rate equation shows the detailed dependence of the reaction rate on the concentrations of reactants and other species present. Different mathematical forms are possible depending on the reaction mechanism. The actual rate equation for a given reaction is determined experimentally and provides information about the reaction mechanism.

Temperature

Temperature usually has a major effect on the rate of a chemical reaction. Molecules at a higher temperature have more thermal energy. Although collision frequency is greater at higher temperatures, this alone contributes only a very small proportion to the increase in rate of reaction. Much more important is the fact that the proportion of reactant molecules with sufficient energy to react (energy greater than activation energy: $E > E_a$) is significantly higher and is explained in detail by the Maxwell–Boltzmann distribution of molecular energies.

The 'rule of thumb' that the rate of chemical reactions doubles for every 10 °C temperature rise is a common misconception. This may have been generalized from the special case of biological systems, where the α (temperature coefficient) is often between 1.5 and 2.5.

A reaction›s kinetics can also be studied with a temperature jump approach. This involves using a sharp rise in temperature and observing the relaxation time of the return to equilibrium. A particularly useful form of temperature jump apparatus is a shock tube, which can rapidly jump a gas›s temperature by more than 1000 degrees.

Catalysts

A catalyst is a substance that alters the rate of a chemical reaction but remains chemically unchanged afterwards. The catalyst increases the rate of the reaction by providing a different reaction mechanism to occur with a lower activation energy. In autocatalysis a reaction product is itself a catalyst for that reaction leading to positive feedback. Proteins that act as catalysts in biochemical reactions are called enzymes. Michaelis–Menten kinetics describe the rate of enzyme mediated reactions. A catalyst does not affect the position of the equilibrium, as the catalyst speeds up the backward and forward reactions equally.

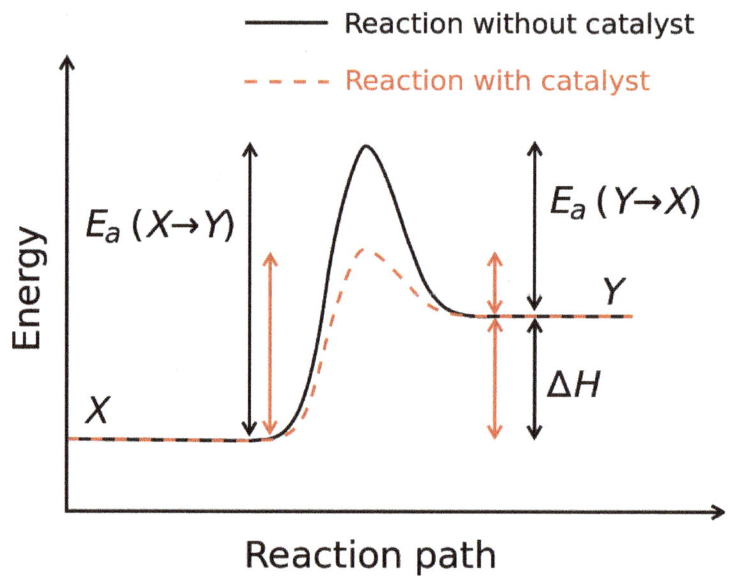

Generic potential energy diagram showing the effect of a catalyst in a hypothetical endothermic chemical reaction. The presence of the catalyst opens a different reaction pathway (shown in red) with a lower activation energy. The final result and the overall thermodynamics are the same.

In certain organic molecules, specific substituents can have an influence on reaction rate in neighbouring group participation.

Pressure

Increasing the pressure in a gaseous reaction will increase the number of collisions between reactants, increasing the rate of reaction. This is because the activity of a gas is directly proportional to the partial pressure of the gas. This is similar to the effect of increasing the concentration of a solution.

In addition to this straightforward mass-action effect, the rate coefficients themselves can change due to pressure. The rate coefficients and products of many high-temperature gas-phase reactions change if an inert gas is added to the mixture; variations on this effect are called fall-off and chemical activation. These phenomena are due to exothermic or endothermic reactions occurring faster than heat transfer, causing the reacting molecules to have non-thermal energy distributions (non-Boltzmann distribution). Increasing the pressure increases the heat transfer rate between the reacting molecules and the rest of the system, reducing this effect.

Condensed-phase rate coefficients can also be affected by (very high) pressure; this is a completely different effect than fall-off or chemical-activation. It is often studied using diamond anvils.

A reaction's kinetics can also be studied with a pressure jump approach. This involves making fast changes in pressure and observing the relaxation time of the return to equilibrium.

Experimental Methods

The experimental determination of reaction rates involves measuring how the concentrations of reactants or products change over time. For example, the concentration of a reactant can be measured by spectrophotometry at a wavelength where no other reactant or product in the system absorbs light.

For reactions which take at least several minutes, it is possible to start the observations after the reactants have been mixed at the temperature of interest.

Fast Reactions

For faster reactions, the time required to mix the reactants and bring them to a specified temperature may be comparable or longer than the half-life of the reaction. Special methods to start fast reactions without slow mixing step include

- Stopped flow methods, which can reduce the mixing time to the order of a millisecond
- Chemical relaxation methods such as temperature jump and pressure jump, in which a pre-mixed system initially at equilibrium is perturbed by rapid heating or depressurization so that it is no longer at equilibrium, and the relaxation back to equilibrium is observed. For example, this method has been used to study the neutralization $H_3O^+ + OH^-$ with a half-life of 1 μs or less under ordinary conditions.
- Flash photolysis, in which a laser pulse produces highly excited species such as free radicals, whose reactions are then studied.

Equilibrium

While chemical kinetics is concerned with the rate of a chemical reaction, thermodynamics determines the extent to which reactions occur. In a reversible reaction, chemical equilibrium is reached when the rates of the forward and reverse reactions are equal (the principle of detailed balance) and the concentrations of the reactants and products no longer change. This is demonstrated by, for example, the Haber–Bosch process for combining nitrogen and hydrogen to produce ammonia. Chemical clock reactions such as the Belousov–Zhabotinsky reaction demonstrate that component concentrations can oscillate for a long time before finally attaining the equilibrium.

Free Energy

In general terms, the free energy change (ΔG) of a reaction determines whether a chemical change will take place, but kinetics describes how fast the reaction is. A reaction can be very exothermic and have a very positive entropy change but will not happen in practice if the reaction is too slow. If a reactant can produce two different products, the thermodynamically most stable one will in general form, except in special circumstances when the reaction is said to be under kinetic reaction control. The Curtin–Hammett principle applies when determining the product ratio for two reactants interconverting rapidly, each going to a different product. It is possible to make predictions about reaction rate constants for a reaction from free-energy relationships.

The kinetic isotope effect is the difference in the rate of a chemical reaction when an atom in one of the reactants is replaced by one of its isotopes.

Chemical kinetics provides information on residence time and heat transfer in a chemical reactor in chemical engineering and the molar mass distribution in polymer chemistry.

Applications and Models

The mathematical models that describe chemical reaction kinetics provide chemists and chemical engineers with tools to better understand and describe chemical processes such as food decomposition, microorganism growth, stratospheric ozone decomposition, and the complex chemistry of biological systems. These models can also be used in the design or modification of chemical reactors to optimize product yield, more efficiently separate products, and eliminate environmentally harmful by-products. When performing catalytic cracking of heavy hydrocarbons into gasoline and light gas, for example, kinetic models can be used to find the temperature and pressure at which the highest yield of heavy hydrocarbons into gasoline will occur.

Chemical Kinetics is frequently validated and explored through modeling in specialized packages as a function of ordinary differential equation-solving (ODE-solving) and curve-fitting.

CHEMICAL PROCESS MODELING

Chemical process modeling is a computer modeling technique used in chemical engineering process design. It typically involves using purpose-built software to define a system of interconnected components, which are then solved so that the steady-state or dynamic behavior of the system can be predicted. The system components and connections are represented as a Process Flow diagram. Simulations can be as simple as the mixing of two substances in a tank, or as complex as an entire alumina refinery.

Chemical process modeling requires a knowledge of the properties of the chemicals involved in the simulation, as well as the physical properties and characteristics of the components of the system, such as tanks, pumps, pipes, pressure vessels, and so on.

PROCESS SIMULATION

Process simulation is used for the design, development, analysis, and optimization of technical processes such as: chemical plants, chemical processes, environmental systems, power stations, complex manufacturing operations, biological processes, and similar technical functions.

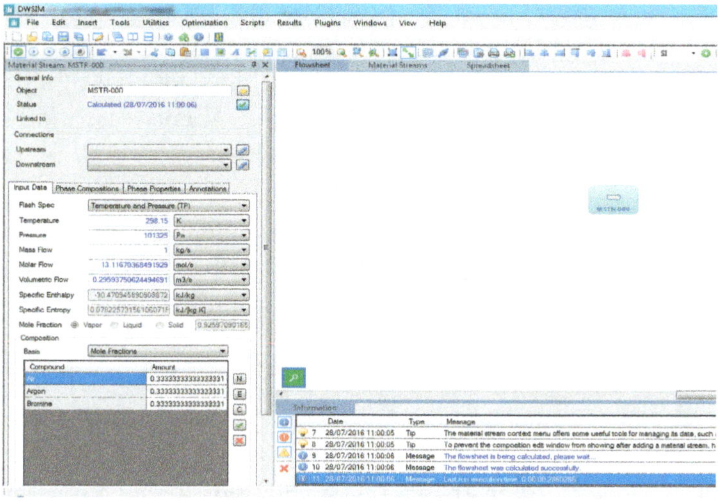

Screenshot of a process simulation software (DWSIM).

Main principle

Process simulation is a **model**-based representation of chemical, physical, biological, and other technical processes and unit operations in software. Basic prerequisites are a thorough knowledge of chemical and physical properties of pure components and mixtures, of reactions, and of mathematical models which, in combination, allow the calculation of a process in computers.

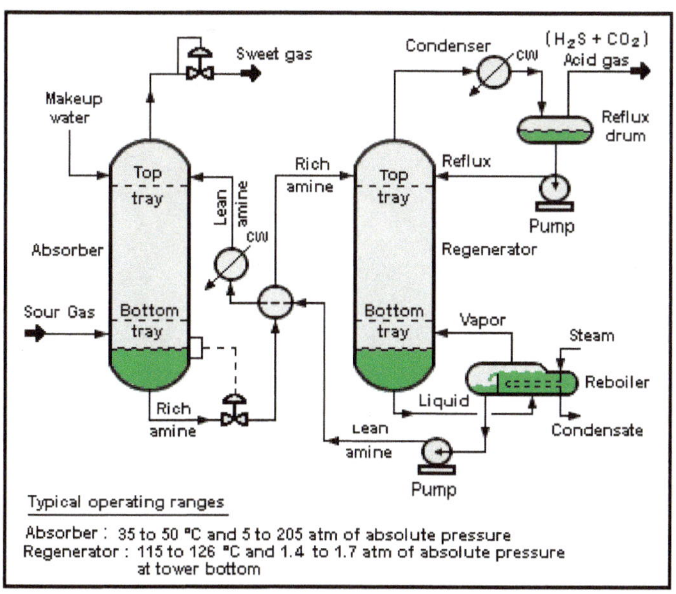

Process flow diagram of a typical amine treating process used in industrial plants

Process simulation software describes processes in flow diagrams where unit operations are positioned and connected by product or educt streams. The software has to solve the mass and energy balance to find a stable operating point. The goal of a process simulation is to find optimal conditions for an examined process. This is essentially an optimization problem which has to be solved in an iterative process.

Process simulation always use models which introduce approximations and assumptions but allow the description of a property over a wide range of temperatures and pressures which might not be covered by real data. Models also allow interpolation and extrapolation - within certain limits - and enable the search for conditions outside the range of known properties.

Modelling

The development of models for a better representation of real processes is the core of the further development of the simulation software. Model development is done on the chemical engineering side but also in control engineering and for the improvement of mathematical simulation techniques. Process simulation is therefore one of the few fields where scientists from chemistry, physics, computer science, mathematics, and several engineering fields work together.

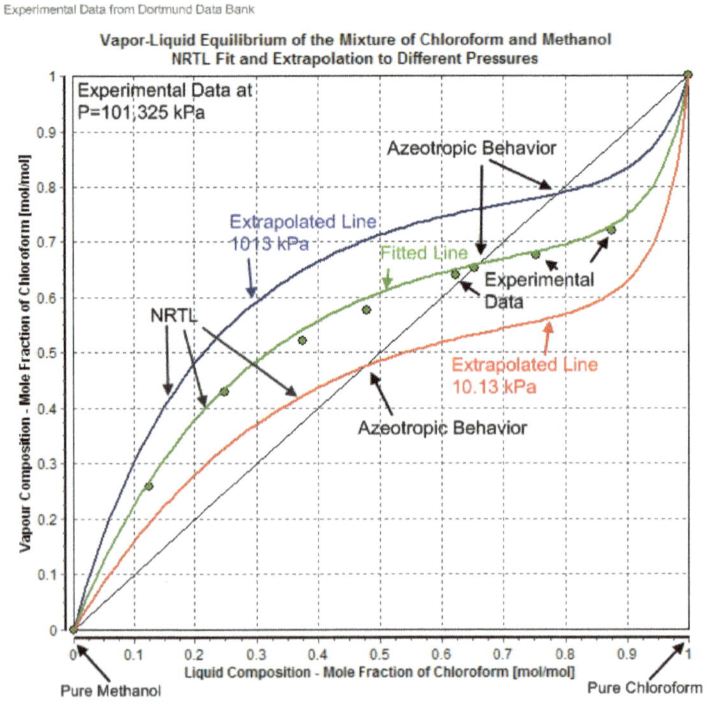

VLE of the mixture of Chloroform and Methanol plus NRTL fit and extrapolation to different pressures.

A lot of efforts are made to develop new and improved models for the calculation of properties. This includes for example the description of

- thermophysical properties like vapor pressures, viscosities, caloric data, etc. of pure components and mixtures
- properties of different apparatuses like reactors, distillation columns, pumps, etc.
- chemical reactions and kinetics
- environmental and safety-related data

Two main different types of models can be distinguished:

- Rather simple equations and correlations where parameters are fitted to experimental data.
- Predictive methods where properties are estimated.

The equations and correlations are normally preferred because they describe the property (almost) exactly. To obtain reliable parameters it is necessary to have experimental data which are usually obtained from factual data banks or, if no data are publicly available, from measurements.

Using predictive methods is much cheaper than experimental work and also than data from data banks. Despite this big advantage predicted properties are normally only used in early steps of the process development to find first approximate solutions and to exclude wrong pathways because these estimation methods normally introduce higher errors than correlations obtained from real data.

Process simulation also encouraged the further development of mathematical models in the fields of numerics and the solving of complex problems.

Steady State and Dynamic Process Simulation

Initially process simulation was used to simulate steady state processes. Steady-state models perform a mass and energy balance of a stationary process (a process in an equilibrium state) it does not depend on time.

Dynamic simulation is an extension of steady-state process simulation whereby time-dependence is built into the models via derivative terms i.e. accumulation of mass and energy. The advent of dynamic simulation means that the time-dependent description, prediction and control of real processes in real time has become possible. This includes the description of starting up and shutting down a plant, changes of conditions during a reaction, holdups, thermal changes and more.

Dynamic simulations require increased calculation time and are mathematically more complex than a steady state simulation. It can be seen as a multiply repeated steady state simulation (based on a fixed time step) with constantly changing parameters.

Dynamic simulation can be used in both an online and offline fashion. The online case being model predictive control, where the real-time simulation results are used to predict the changes that would occur for a control input change, and the control parameters are optimized based on the results. Offline process simulation can be used in the design, troubleshooting and optimization of process plant as well as the conduction of case studies to assess the impacts of process modifications. Dynamic simulation is also used for operator training.

PROCESS OPTIMIZATION

Process optimization is the discipline of adjusting a process so as to optimize some specified set of parameters without violating some constraint. The most common goals are minimizing cost and maximizing throughput and/or efficiency. This is one of the major quantitative tools in industrial decision making.

When optimizing a process, the goal is to maximize one or more of the process specifications, while keeping all others within their constraints. This can be done by using a process mining tool, discovering the critical activities and bottlenecks, and acting only on them.

Areas

Fundamentally, there are three parameters that can be adjusted to affect optimal performance. They are:

- Equipment optimization

The first step is to verify that the existing equipment is being used to its fullest advantage by examining operating data to identify equipment bottlenecks.

- Operating procedures

Operating procedures may vary widely from person-to-person or from shift-to-shift. Automation of the plant can help significantly. But automation will be of no help if the operators take control and run the plant in manual.

- Control optimization

In a typical processing plant, such as a chemical plant or oil refinery, there are hundreds or even thousands of control loops. Each control loop is responsible for controlling one part of the process, such as maintaining a temperature, level, or flow.

If the control loop is not properly designed and tuned, the process runs below its optimum. The process will be more expensive to operate, and equipment will wear out prematurely. For each control loop to run optimally, identification of sensor, valve, and tuning problems is important. It has been well documented that over 35% of control loops typically have problems.

The process of continuously monitoring and optimizing the entire plant is sometimes called performance supervision.

REFERENCES

1. ISO Online Browsing Platform (OBP)-ISO 14040:2006(en)
2. C.M. Guldberg and P. Waage,"Studies Concerning Affinity" C. M. Forhandlinger: Videnskabs-Selskabet i Christiana (1864), 35

3. P. Waage, “Experiments for Determining the Affinity Law” ,Forhandlinger i Videnskabs-Selskabet i Christiania, (1864) 92.
4. C.M. Guldberg, “Concerning the Laws of Chemical Affinity”, C. M. Forhandlinger i Videnskabs-Selskabet i Christiania (1864) 111
5. Hoff, J. H. van’t (Jacobus Henricus van’t); Cohen, Ernst; Ewan, Thomas (1896-01-01). Studies in chemical dynamics. Amsterdam : F. Muller ; London : Williams & Norgate.
6. The Nobel Prize in Chemistry 1901, Nobel Prizes and Laureates, official website.
7. A.N. Gorban, G.S. Yablonsky Three Waves of Chemical Dynamics, Mathematical Modelling of Natural Phenomena 10(5) (2015), p. 1–5.
8. Laidler, K. J. Chemical Kinetics (3rd ed., Harper and Row 1987) p.33-39 ISBN 0-06-043862-2
9. Espenson, J.H. Chemical Kinetics and Reaction Mechanisms (2nd ed., McGraw-Hill 2002), p.254-256 ISBN 0-07-288362-6
10. Atkins P. and de Paula J., Physical Chemistry (8th ed., W.H. Freeman 2006) p.793 ISBN 0-7167-8759-8
11. Espenson,J.H. Chemical Kinetics and Reaction Mechanisms (2nd ed., McGraw-Hill 2002), p.256-8 ISBN 0-07-288362-6
12. Steinfeld J.I., Francisco J.S. and Hase W.L. Chemical Kinetics and Dynamics (2nd ed., Prentice-Hall 1999) p.140-3 ISBN 0-13-737123-3
13. Atkins P. and de Paula J., Physical Chemistry (8th ed., W.H. Freeman 2006) pp.805-7 ISBN 0-7167-8759-8
14. Laidler, K.J. Chemical Kinetics (3rd ed., Harper and Row 1987) p.359-360 ISBN 0-06-043862-2
15. Espenson, J.H. Chemical Kinetics and Reaction Mechanisms (2nd ed., McGraw-Hill 2002), p.264-6 ISBN 0-07-288362-6
16. Steinfeld J.I., Francisco J.S. and Hase W.L. Chemical Kinetics and Dynamics (2nd ed., Prentice-Hall 1999) p.94-97 ISBN 0-13-737123-3
17. “Chemical Kinetics: Simple Binding: F + G $\leftrightharpoons$ B” (PDF). Civilized Software, Inc. Retrieved 2015-09-01.

CHAPTER 5

Chemical Process Design

Chapter Contents

- **Introduction**
- **Documentation**
- **Design Considerations**
- **Process Integration**
- **Process Engineering**

Process design is the activity of determining the workflow, equipment needs, and implementation requirements for a particular process. Process design typically uses a number of tools including flowcharting, process simulation software, and scale models.

INTRODUCTION

In Chemical engineering, process design is the design of processes for desired physical and/or chemical transformation of materials. Process design is central to chemical engineering, and it can be considered to be the summit of that field, bringing together all of the field's components.

Process design can be the design of new facilities or it can be the modification or expansion of existing facilities. The design starts at a conceptual level and ultimately ends in the form of fabrication and construction plans.

Process design is distinct from equipment design, which is closer in spirit to the design of unit operations. Processes often include many unit operations.

DOCUMENTATION

Process design documents serve to define the design and they ensure that the design components fit together. They are useful in communicating ideas and plans to other engineers involved with the design, to external regulatory agencies, to equipment vendors and to construction contractors.

In order of increasing detail, process design documents include:

- Block flow diagrams (BFD): Very simple diagrams composed of rectangles and lines indicating major material or energy flows.
- Process flow diagrams (PFD): Typically more complex diagrams of major unit operations as well as flow lines. They usually include a material balance, and sometimes an energy balance, showing typical or design flowrates, stream compositions, and stream and equipment pressures and temperatures.
- Piping and instrumentation diagrams (P&ID): Diagrams showing each and every pipeline with piping class (carbon steel or stainless steel) and pipe size (diameter). They also show valving along with instrument locations and process control schemes.
- Specifications: Written design requirements of all major equipment items.

Process designers also typically write operating manuals on how to start-up, operate and shut-down the process.

Documents are maintained after construction of the process facility for the operating personnel to refer to. The documents also are useful when modifications to the facility are planned.

A primary method of developing the process documents is process flowsheeting.

DESIGN CONSIDERATIONS

There are several considerations that need to be made when designing any chemical process unit. Design conceptualization and considerations can begin once product purities, yields, and throughput rates are all defined.

Objectives that a design may strive to include:

- Throughput rate
- Process yield
- Product purity

Constraints include:

- Capital cost
- Available space
- Safety concerns
- Environmental impact and projected effluents and emissions
- Waste production
- Operating and maintenance costs

Other factors that designers may include are:

- Reliability
- Redundancy
- Flexibility
- Anticipated variability in feedstock and allowable variability in product.

Sources of Design Information

Designers usually do not start from scratch, especially for complex projects. Often the engineers have pilot plant data available or data from full-scale operating facilities. Other sources of information include proprietary design criteria provided by process licensors, published scientific data, laboratory experiments, and input.

Computer help

The advent of low cost powerful computers has aided complex mathematical simulation of processes, and simulation software is often used by design engineers. Simulations can identify weaknesses in designs and allow engineers to choose better alternatives.

However, engineers still rely on heuristics, intuition, and experience when designing a process. Human creativity is an element in complex designs.

PROCESS INTEGRATION

Process integration is a term in chemical engineering which has two possible meanings.

1. A holistic approach to process design which emphasizes the unity of the process and considers the interactions between different unit operations from the outset, rather than optimising them separately. This can also be called *integrated process design* or *process synthesis*. El-Halwagi (1997 and 2006) and Smith (2005) describe the approach well. An important first step is often *product design* (Cussler and Moggridge 2003) which develops the specification for the product to fulfil its required purpose.
2. *Pinch analysis*, a technique for designing a process to minimize energy consumption and maximise heat recovery, also known as *heat integration, energy integration* or *pinch technology*. The technique calculates thermodynamically attainable *energy targets* for a given process and identifies how to achieve them. A key insight is the pinch temperature, which is the most constrained point in the process. The most detailed explanation of the techniques is by Linnhoff et al. (1982), Shenoy (1995) and Kemp (2006). This definition reflects the fact that the first major success for process integration was the thermal pinch analysis addressing energy problems and pioneered by Linnhoff and co-workers. Later, other pinch analyses were developed for several applications such as mass-exchange networks (El-Halwagi and Manousiouthakis, 1989), water minimization (Wang and Smith, 1994), and material recycle (El-Halwagi et al., 2003). A very successful extension was "Hydrogen Pinch", which was applied to

refinery hydrogen management (Nick Hallale et al., 2002 and 2003). This allowed refiners to minimize the capital and operating costs of hydrogen supply to meet ever stricter environmental regulations and also increase hydrotreater yields.

In the context of chemical engineering, Process Integration can be defined as a holistic approach to process design and optimization, which exploits the interactions between different units in order to employ resources effectively and minimize costs.

Note that Process Integration is not limited to the design of new plants, but it also covers retrofit design (e.g. new units to be installed in an old plant) and the operation of existing systems. Nick Hallale (2001), in his article in Chemical Engineering Progress provided a state of the art review. He explained that process integration far wider scope and touches every area of process design. Industries are making more money from their raw materials and capital assets while becoming cleaner and more sustainable.

The main advantage of process integration is to consider a system as a whole (i.e. integrated or holistic approach) in order to improve their design and/or operation. In contrast, an analytical approach would attempt to improve or optimize process units separately without necessarily taking advantage of potential interactions among them.

For instance, by using process integration techniques it might be possible to identify that a process can use the heat rejected by another unit and reduce the overall energy consumption, even if the units are not running at optimum conditions on their own. Such an opportunity would be missed with an analytical approach, as it would seek to optimize each unit, and thereafter it wouldn't be possible to re-use the heat internally.

Typically, process integration techniques are employed at the beginning of a project (e.g. a new plant or the improvement of an existing one) to screen out promising options to optimize the design and/or operation of a process plant.

Also it is often employed, in conjunction with simulation and mathematical optimization tools to identify opportunities in order to better integrate a system (new or existing) and reduce capital and/or operating costs.

Most process integration techniques employ Pinch analysis or Pinch Tools to evaluate several processes as a whole system. Therefore, strictly speaking, both concepts are not the same, even if in certain contexts they are used interchangeably. The review by Nick Hallale (2001) explains that in the future, several trends are to be expected in the field. In the future, it seems probable that the boundary between targets and design will be blurred and that these will be based on more structural information regarding the process network. Second, it is likely that we will see a much wider range of applications of process integration. There is still much work to be carried out in the area of separation, not only in complex distillation systems, but also in mixed types of separation systems. This includes processes involving solids, such as flotation and crystallization. The use of process integration techniques for reactor design has seen rapid progress, but is still in its early stages. Third, a new generation of software tools is expected. The emergence of commercial software for process integration is fundamental to its wider application in process design.

PROCESS ENGINEERING

Process engineering focuses on the design, operation, control, optimization and Intensification of chemical, physical, and biological processes. Process engineering encompasses a vast range of industries, such as chemical, petrochemical, agriculture, mineral processing, advanced material, food, pharmaceutical, software development and biotechnological industries.

Process engineering involves translating the needs of the customer into (typically) production facilities that convert "raw materials" into value-added components that are transported to the next stage of the supply chain, typically "packaging engineering", but some larger volume processes such as petroleum refining tend to transfer the products into transportation (trucks or rail) that are then directed to distributors or bulk outlets. Prior to construction, the design work of process engineering begins with a "block diagram" showing raw materials and the transformations/unit operations desired. The design work then progresses to a Process flow diagram (PFD) where material flow paths, storage equipment (such as tanks and silos), transformations/Unit Operations (such as distillation columns, receiver/head tanks, mixing, separations, pumping, etc.) and flowrates are specified, as well as a list of all pipes and conveyors and their contents, material properties such as density, viscosity, particle size distribution, flow rates, pressures, temperatures, and materials of construction for the piping and unit operations. The process flow diagram is then used to develop a Piping and instrumentation diagram (P&ID) which includes pipe and conveyor sizing information to address the desired flowrates, process controls (such as tank level indications, material flow meters, weighing devices, motor speed controls, temperature and pressure indicators/ controllers, etc.). The P&ID is then used as a basis of design for developing the "system operation guide" or "functional design specification" which outlines the operation of the process. From the "P&ID", a proposed layout (general arrangement) of the process can be shown from an overhead view (plot plan) and a side view (elevation), and other engineering disciplines are involved such as civil engineers for site work (earth moving), foundation design, concrete slab design work, structural steel to support the equipment, etc.). All previous work is directed toward defining the scope of the project, then developing a cost estimate to get the design installed, and a schedule to communicate the timing needs for engineering, procurement, fabrication, installation, commissioning, startup, and ongoing production of the process. Depending on the needed accuracy of the cost estimate and schedule that is required, several iterations of designs are generally provided to customers or stakeholders who feedback their requirements and the process engineer incorporates these additional instructions and wants (scope revisions) into the overall design and additional cost estimates and Schedules are developed for funding approval. Following funding approval, the project is executed via project management.

The application of systematic computer-based methods to process engineering is process systems engineering.

Significant Accomplishments

Several accomplishments have been made in Process Systems Engineering:

- Process design: synthesis of energy recovery networks, synthesis of distillation systems (azeotropic), synthesis of reactor networks, hierarchical decomposition flowsheets, superstructure optimization, design multiproduct batch plants. Design of the production reactors for the production of plutonium, design of nuclear submarines.

- Process control: model predictive control, controllability measures, robust control, nonlinear control, statistical process control, process monitoring, thermodynamics-based control
- Process operations: scheduling process networks, multiperiod planning and optimization, data reconciliation, real-time optimization, flexibility measures, fault diagnosis
- Supporting tools: sequential modular simulation, equation based process simulation, AI/expert systems, large-scale nonlinear programming (NLP), optimization of differential algebraic equations (DAEs), mixed-integer nonlinear programming (MINLP), global optimization

History of Process Systems Engineering

Process systems engineering (PSE) is a relatively young area in chemical engineering. The first time that this term was used was in a Special Volume of the AIChE Symposium Series in 1961. However, it was not until 1982 when the first international symposium on this topic took place in Kyoto, Japan, that the term PSE started to become widely accepted.

The first textbook in the area was "Strategy of Process Engineering" by Dale F. Rudd and Charles C. Watson, Wiley, 1968. The Computing and Systems Technology (CAST) Division, Area 10 of AIChE, was founded in 1977 and currently has about 1200 members. CAST has four sections: Process Design, Process Control, Process Operations, and Applied Mathematics.

The first journal devoted to PSE was "Computers and Chemical Engineering," which appeared in 1977. The Foundations of Computer-Aided Process Design (FOCAPD) conference in 1980 in Henniker was one of the first meetings in a series on that topic in the PSE area. It is now accompanied by the successful series on Control (CPC), Operations (FOCAPO), and the world-wide series entitled Process Systems Engineering. The CACHE Corporation (Computer Aids for Chemical Engineering), which organizes these conferences, was initially launched by academics in 1970, motivated by the introduction of process simulation in the chemical engineering curriculum.

Roger W.H. Sargent from Imperial College was one of the pioneers in the area. PSE is an active area of research in many other countries, particularly in the United Kingdom, Germany, Japan, Korea, and China.

REFERENCES

1. http://www.allbusiness.com/manufacturing/chemical-manufacturing/1128450-1.html#ixzz1Wij9OVl5

2. Cussler, E.L. and Moggridge, G.D. (2001). Chemical Product Design. Cambridge University Press (Cambridge Series in Chemical Engineering). ISBN 0-521-79183-9
3. El-Halwagi, M. M., (1997) "Pollution Prevention through Process Integration", Academic Press
4. El-Halwagi, M. M., F. Gabriel, and D. Harell, (2003) "Rigorous Graphical Targeting for Resource Conservation via Material Recycle/Reuse Networks", Ind. Eng. Chem. Res., 42, 4319-4328
5. El-Halwagi, M. M., and Manousiouthakis, V. (1989). Synthesis of mass exchange networks. AIChE J. 35(8), 1233-1244.
6. Hallale, Nick, (2001), "Burning Bright: Trends in Process Integration", Chemical Engineering Progress, July 2001
7. Nick Hallale, I Moore, D. Vauk (2002), "Hydrogen: Liability or Asset?", Chemical Engineering Progress, September 2002
8. Hallale, N. Ian Moore, Dennis Vauk, "Hydrogen optimization at minimal investment", Petroleum Technology Quarterly (PTQ), Spring (2003)
9. Kemp, I.C. (2006). Pinch Analysis and Process Integration: A User Guide on Process Integration for the Efficient Use of Energy, 2nd edition. Butterworth-Heinemann. ISBN 0-7506-8260-4. Includes downloadable spreadsheet software.
10. Linnhoff, B., D.W. Townsend, D. Boland, G.F. Hewitt, B.E.A. Thomas, A.R. Guy and R.H. Marsland, (1982) "A User Guide on Process Integration for the Efficient Use of Energy," IChemE UK
11. Shenoy, U.V. (1995). "Heat Exchanger Network Synthesis: Process Optimization by Energy and Resource Analysis". Includes two computer disks. Gulf Publishing Company, Houston, TX, USA. ISBN 0-88415-391-6.
12. Smith, R. (2005). Chemical Process Design and Integration. John Wiley and Sons. ISBN 0-471-48680-9
13. Wang, Y. P. and R. Smith (1994). Wastewater Minimisation. Chem. Eng. Sci., 49, 981-1006
14. Research Challenges in Process Systems Engineering by Ignacio E. Grossmann and Arthur W. Westerberg, Department of Chemical Engineering at Carnegie Mellon University in Pittsburgh, PA

Chemical Process

Chapter Contents

A chemical process is any laboratory or industrial activity where products are chemically reacted with each other to form new products. The reactions change the structure of the product molecules, rather than adding them together in a mixing process. Mixtures can result from these activities, but a chemical change has occurred in one or more of the reacted materials.

Process designs often involve building larger molecules from smaller molecules found in raw materials. Some reactions can break larger molecules and rearrange them in different ways to produce a final product. Multiple reactions may be required to produce a desired molecule, and can include additional raw materials or changes in temperature or pressure to cause different products to form. A process may be a single reaction step, or an industrial operation creating finished chemical products on a large scale.

Chemicals are first produced in laboratory-scale equipment, both to confirm the necessary steps and to fine-tune the reaction temperatures and pressures often required for the reactions to occur. Many industrial chemical processes occur at high temperatures and/or pressures, but less often, the reactions can occur at very low pressures or in a vacuum, or at very low temperatures.

INTRODUCTION

In a scientific sense, a chemical process is a method or means of somehow changing one or more chemicals or chemical compounds. Such a chemical process can occur by itself or be caused by an outside force, and involves a chemical reaction of some sort. In an "engineering" sense, a chemical process is a method intended to be used in manufacturing or on an industrial scale to change the composition of chemical(s) or material(s), usually using technology similar or related to that used in chemical plants or the chemical industry.

Neither of these definitions is exact in the sense that one can always tell definitively what is a chemical process and what is not; they are practical definitions. There is also significant overlap in these two definition variations. Because of the inexactness of the definition, chemists and other scientists use the term "chemical process" only in a general sense or in the engineering sense. However, in the "process (engineering)" sense, the term "chemical process" is used extensively.

Although this type of chemical process may sometimes involve only one step, often multiple steps, referred to as unit operations, are involved. In a plant, each of the unit operations commonly occur in individual vessels or sections of the plant called units. Often, one or more chemical reactions are involved, but other ways of changing chemical (or material) composition may be used, such as mixing or separation processes. The process steps may be sequential in time or sequential in space along a stream of flowing or moving material. For a given amount of a feed (input) material or product (output) material, an expected amount of material can be determined at key steps in the process from empirical data and material balance calculations. These amounts can be scaled up or down to suit the desired capacity or operation of a particular chemical plant built for such a process. More than one chemical plant may use the same chemical process, each plant perhaps at differently scaled capacities. Chemical processes like distillation and crystallization go back to alchemy in Alexandria, Egypt.

Such chemical processes can be illustrated generally as block flow diagrams or in more detail as process flow diagrams. Block flow diagrams show the units as blocks and the streams flowing between them as connecting lines with arrowheads to show direction of flow.

In addition to chemical plants for producing chemicals, chemical processes with similar technology and equipment are also used in oil refining and other refineries, natural gas processing, polymer and pharmaceutical manufacturing, food processing, and water and wastewater treatment.

Unit Processing in Chemical Process

Unit processing is the basic processing in chemical engineering. Together with unit operations it forms the main principle of the varied chemical industries. Each genre of unit processing follows the same chemical law much as each genre of unit operations follows the same physical law.

Chemical engineering unit processing consists of the following important processes:

- Oxidation

- Reduction
- Hydrogenation
- Dehydrogenation
- Hydrolysis
- Hydration
- Dehydration
- Halogenation
- Nitrification
- Sulfonation
- Ammoniation
- Alkaline fusion
- Alkylation
- Dealkylation
- Esterification
- Polymerization
- Polycondensation
- Catalysis

REDOX

Redox (short for reduction–oxidation reaction) is a chemical reaction in which the oxidation states of atoms are changed. Any such reaction involves both a reduction process and a complementary oxidation process, two key concepts involved with electron transfer processes. Redox reactions include all chemical reactions in which atoms have their oxidation state changed; in general, redox reactions involve the transfer of electrons between chemical species. The chemical species from which the electron is stripped is said to have been oxidized, while the chemical species to which the electron is added is said to have been reduced. It can be explained in simple terms:

- Oxidation is the loss of electrons or an increase in oxidation state by a molecule, atom, or ion.
- Reduction is the gain of electrons or a decrease in oxidation state by a molecule, atom, or ion.

As an example, during the combustion of wood, oxygen from the air is reduced, transferring electrons from the carbon. Although oxidation reactions are commonly associated with the formation of oxides from oxygen molecules, oxygen is not necessarily included in such reactions, as other chemical species can serve the same function.

The reaction can occur relatively slowly, as in the case of rust, or more quickly, as in the case of fire. There are simple redox processes, such as the oxidation of carbon to yield carbon dioxide (CO_2) or the reduction of carbon by hydrogen to yield methane (CH_4), and more complex processes such as the oxidation of glucose ($C_6H_{12}O_6$) in the human body.

Rusting iron.

A bonfire.

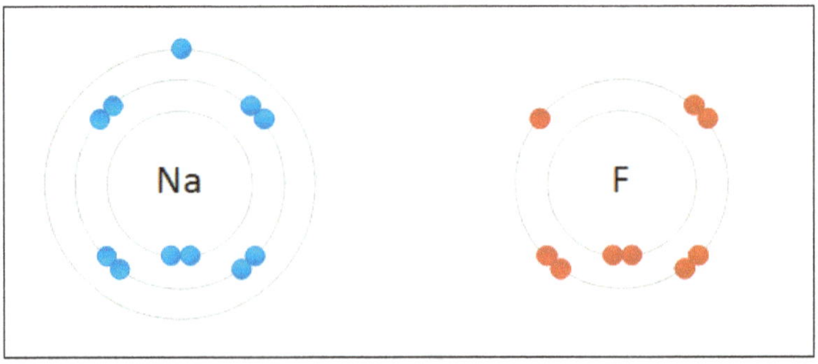

Sodium and fluorine bonding ionically to form sodium fluoride. Sodium loses its outer electron to give it a stable electron configuration, and this electron enters the fluorine atom exothermically. The oppositely charged ions are then attracted to each other. The sodium is oxidized, and the fluorine is reduced.

Etymology

"Redox" is a combination of "reduction" and "oxidation".

The word oxidation originally implied reaction with oxygen to form an oxide, since dioxygen (O_2) was historically the first recognized oxidizing agent. Later, the term was expanded to encompass oxygen-like substances that accomplished parallel chemical reactions. Ultimately, the meaning was generalized to include all processes involving loss of electrons.

The word reduction originally referred to the loss in weight upon heating a metallic ore such as a metal oxide to extract the metal. In other words, ore was "reduced" to metal. Antoine Lavoisier (1743–1794) showed that this loss of weight was due to the loss of oxygen as a gas. Later, scientists realized that the metal atom gains electrons in this process. The meaning of reduction then became generalized to include all processes involving gain of electrons. Even though "reduction" seems counter-intuitive when speaking of the gain of electrons, it might help to think of reduction as the loss of oxygen, which was its historical meaning. Since electrons are negatively charged, it is also helpful to think of this as reduction in electrical charge.

The electrochemist John Bockris has used the words electronation and deelectronation to describe reduction and oxidation processes respectively when they occur at electrodes. These words are analogous to protonation and deprotonation, but they have not been widely adopted by chemists.

The term "hydrogenation" could be used instead of reduction, since hydrogen is the reducing agent in a large number of reactions, especially in organic chemistry and biochemistry. But, unlike oxidation, which has been generalized beyond its root element, hydrogenation has maintained its specific connection to reactions that add hydrogen to another substance (e.g., the hydrogenation of unsaturated fats into saturated fats, $R{-}CH{=}CH{-}R + H_2 \rightarrow R{-}CH_2{-}CH_2{-}R$). The word "redox" was first used in 1928.

Definitions

The processes of oxidation and reduction occur simultaneously and cannot happen independently of one another, similar to the acid–base reaction. The oxidation alone and the reduction alone are each called a half-reaction, because two half-reactions always occur together to form a whole reaction. When writing half-reactions, the gained or lost electrons are typically included explicitly in order that the half-reaction be balanced with respect to electric charge.

Though sufficient for many purposes, these general descriptions are not precisely correct. Although oxidation and reduction properly refer to a change in oxidation state — the actual transfer of electrons may never occur. The oxidation state of an atom is the fictitious charge that an atom would have if all bonds between atoms of different elements were 100% ionic. Thus, oxidation is best defined as an increase in oxidation state, and reduction as a decrease in oxidation state. In practice, the transfer of electrons will always cause a change in oxidation state, but there are many reactions that are classed as "redox" even though no electron transfer occurs (such as those involving covalent bonds).

Oxidizing and Reducing Agents

In redox processes, the reductant transfers electrons to the oxidant. Thus, in the reaction, the reductant or reducing agent loses electrons and is oxidized, and the oxidant or oxidizing agent gains electrons and is reduced. The pair of an oxidizing and reducing agent that are involved in a particular reaction is called a redox pair. A redox couple is a reducing species and its corresponding oxidizing form, e.g., Fe^{2+}/Fe^{3+}.

Oxidizers

Substances that have the ability to oxidize other substances (cause them to lose electrons) are said to be oxidative or oxidizing and are known as oxidizing agents, oxidants, or oxidizers. That is, the oxidant (oxidizing agent) removes electrons from another substance, and is thus itself reduced. And, because it "accepts" electrons, the oxidizing agent is also called an electron acceptor. Oxygen is the quintessential oxidizer.

The international pictogram for oxidising chemicals.

Oxidants are usually chemical substances with elements in high oxidation states (e.g., H_2O_2, MnO^{-4}, CrO_3, $Cr_2O_2^{-7}$, OsO_4), or else highly electronegative elements (O_2, F_2, Cl_2, Br_2) that can gain extra electrons by oxidizing another substance.

Reducers

Substances that have the ability to reduce other substances (cause them to gain electrons) are said to be reductive or reducing and are known as reducing agents, reductants, or reducers. The reductant (reducing agent) transfers electrons to another substance, and is thus itself oxidized. And, because it "donates" electrons, the reducing agent is also called an electron donor. Electron donors can also form charge transfer complexes with electron acceptors.

Reductants in chemistry are very diverse. Electropositive elemental metals, such as lithium, sodium, magnesium, iron, zinc, and aluminium, are good reducing agents. These metals donate or give away electrons readily. Hydride transfer reagents, such as $NaBH_4$ and $LiAlH_4$, are widely used in organic chemistry, primarily in the reduction of carbonyl compounds to alcohols. Another method of reduction involves the use of hydrogen gas (H_2) with a palladium, platinum, or nickel catalyst. These catalytic reductions are used primarily in the reduction of carbon-carbon double or triple bonds.

Standard Electrode Potentials (Reduction Potentials)

Each half-reaction has a standard electrode potential (E^0_{cell}), which is equal to the potential difference or voltage at equilibrium under standard conditions of an electrochemical cell in which the cathode reaction is the half-reaction considered, and the anode is a standard hydrogen electrode where hydrogen is oxidized:

$\frac{1}{2} H_2 \rightarrow H^+ + e^-$.

The electrode potential of each half-reaction is also known as its reduction potential E^0_{red}, or potential when the half-reaction takes place at a cathode. The reduction potential is a measure of the tendency of the oxidizing agent to be reduced. Its value is zero for $H^+ + e^- \rightarrow \frac{1}{2} H_2$ by definition, positive for oxidizing agents stronger than H^+ (e.g., +2.866 V for F_2) and negative for oxidizing agents that are weaker than H^+ (e.g., −0.763 V for Zn^{2+}).

For a redox reaction that takes place in a cell, the potential difference is:

$$E^0_{cell} = E^0_{cathode} - E^0_{anode}$$

However, the potential of the reaction at the anode was sometimes expressed as an oxidation potential:

$$E^0_{ox} = -E^0_{red}.$$

The oxidation potential is a measure of the tendency of the reducing agent to be oxidized, but does not represent the physical potential at an electrode. With this notation, the cell voltage equation is written with a plus sign

$$E^0_{cell} = E^0_{red(cathode)} + E^0_{ox(anode)}$$

Examples of Redox Reactions

A good example is the reaction between hydrogen and fluorine in which hydrogen is being oxidized and fluorine is being reduced:

$H_2 + F_2 \rightarrow 2\ HF$

We can write this overall reaction as two half-reactions:

the oxidation reaction:

$H_2 \rightarrow 2\ H^+ + 2\ e^-$

and the reduction reaction:

$F_2 + 2\ e^- \rightarrow 2\ F^-$

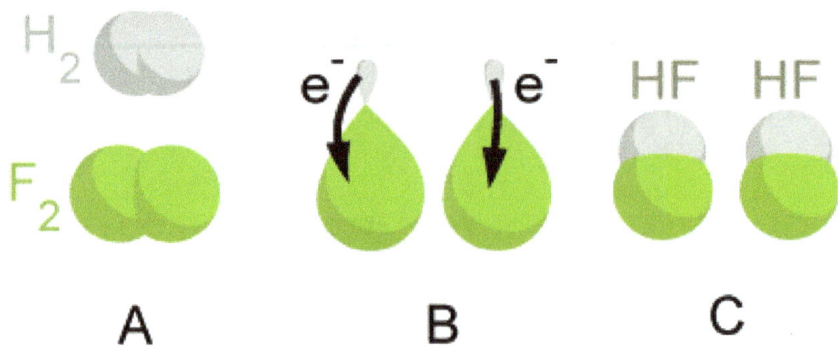

Illustration of a redox reaction.

Analyzing each half-reaction in isolation can often make the overall chemical process clearer. Because there is no net change in charge during a redox reaction, the number of electrons in excess in the oxidation reaction must equal the number consumed by the reduction reaction (as shown above).

Elements, even in molecular form, always have an oxidation state of zero. In the first half-reaction, hydrogen is oxidized from an oxidation state of zero to an oxidation state of +1. In the second half-reaction, fluorine is reduced from an oxidation state of zero to an oxidation state of −1.

When adding the reactions together the electrons are canceled:

$$H_2 \rightarrow 2\ H^+ + 2\ e^-$$

$$F_2 + 2\ e^- \rightarrow 2\ F^-$$

$$H_2 + F_2 \rightarrow 2\ H^+ + 2\ F^-$$

And the ions combine to form hydrogen fluoride:

$2\ H^+ + 2\ F^- \rightarrow 2\ HF$

The overall reaction is:

$H_2 + F_2 \rightarrow 2\ HF$

Metal Displacement

In this type of reaction, a metal atom in a compound (or in a solution) is replaced by an atom of another metal. For example, copper is deposited when zinc metal is placed in a copper(II) sulfate solution:

$Zn(s) + CuSO_4(aq) \rightarrow ZnSO_4(aq) + Cu(s)$

In the above reaction, zinc metal displaces the copper(II) ion from copper sulfate solution and thus liberates free copper metal.

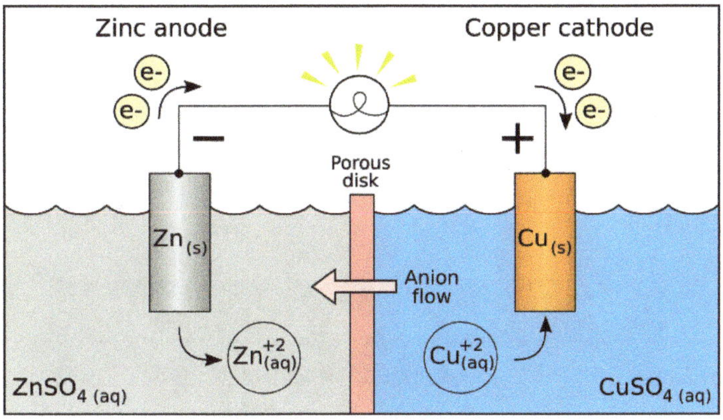

A redox reaction is the force behind an electrochemical cell like the Galvanic cell pictured. The battery is made out of a zinc electrode in a $ZnSO_4$ solution connected with a wire and a porous disk to a copper electrode in a $CuSO_4$ solution.

The ionic equation for this reaction is:

$Zn + Cu^{2+} \rightarrow Zn^{2+} + Cu$

As two half-reactions, it is seen that the zinc is oxidized:

$Zn \rightarrow Zn^{2+} + 2\ e^-$

And the copper is reduced:

$Cu^{2+} + 2\ e^- \rightarrow Cu$

Other Examples

- The reduction of nitrate to nitrogen in the presence of an acid (denitrification):

  $$2\ NO^{-3} + 10\ e^- + 12\ H^+ \rightarrow N_2 + 6\ H_2O$$

- The combustion of hydrocarbons, such as in an internal combustion engine, which produces water, carbon dioxide, some partially oxidized forms such as carbon monoxide, and heat energy. Complete oxidation of materials containing carbon produces carbon dioxide.
- In organic chemistry, the stepwise oxidation of a hydrocarbon by oxygen produces water and, successively, an alcohol, an aldehyde or a ketone, a carboxylic acid, and then a peroxide.

Corrosion and Rusting

The term corrosion refers to the electrochemical oxidation of metals in reaction with an oxidant such as oxygen. Rusting, the formation of iron oxides, is a well-known example of electrochemical corrosion; it forms as a result of the oxidation of iron metal. Common rust often refers to iron(III) oxide, formed in the following chemical reaction:

- $4\ Fe + 3\ O_2 \rightarrow 2\ Fe_2O_3$
- The oxidation of iron(II) to iron(III) by hydrogen peroxide in the presence of an acid:

$Fe^{2+} \rightarrow Fe^{3+} + e^-$

$H_2O_2 + 2\ e^- \rightarrow 2\ OH^-$

Overall equation:

$2\ Fe^{2+} + H_2O_2 + 2\ H^+ \rightarrow 2\ Fe^{3+} + 2\ H_2O$

Oxides, such as iron(III) oxide or rust, which consists of hydrated iron(III) oxides $Fe_2O_3{\cdot}nH_2O$ and iron(III) oxide-hydroxide (FeO(OH), $Fe(OH)_3$), form when oxygen combines with other elements.

Iron rusting in pyrite cubes.

Redox Reactions in Industry

Cathodic protection is a technique used to control the corrosion of a metal surface by making it the cathode of an electrochemical cell. A simple method of protection connects protected metal to a more easily corroded "sacrificial anode" to act as the anode. The sacrificial metal instead of the protected metal, then, corrodes. A common application of cathodic protection is in galvanized steel, in which a sacrificial coating of zinc on steel parts protects them from rust.

The primary process of reducing ore at high temperature to produce metals is known as smelting.

Oxidation is used in a wide variety of industries such as in the production of cleaning products and oxidizing ammonia to produce nitric acid, which is used in most fertilizers.

Redox reactions are the foundation of electrochemical cells, which can generate electrical energy or support electrosynthesis.

The process of electroplating uses redox reactions to coat objects with a thin layer of a material, as in chrome-platedautomotive parts, silver plating cutlery, and gold-plated jewelry.

The production of compact discs depends on a redox reaction, which coats the disc with a thin layer of metal film.

Redox Reactions in Biology

Many important biological processes involve redox reactions.

Cellular respiration, for instance, is the oxidation of glucose ($C_6H_{12}O_6$) to CO_2 and the reduction of oxygen to water. The summary equation for cell respiration is:

$C_6H_{12}O_6 + 6\ O_2 \rightarrow 6\ CO_2 + 6\ H_2O$

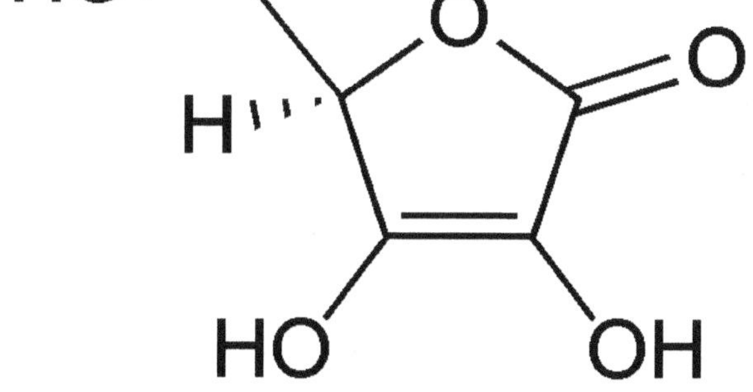

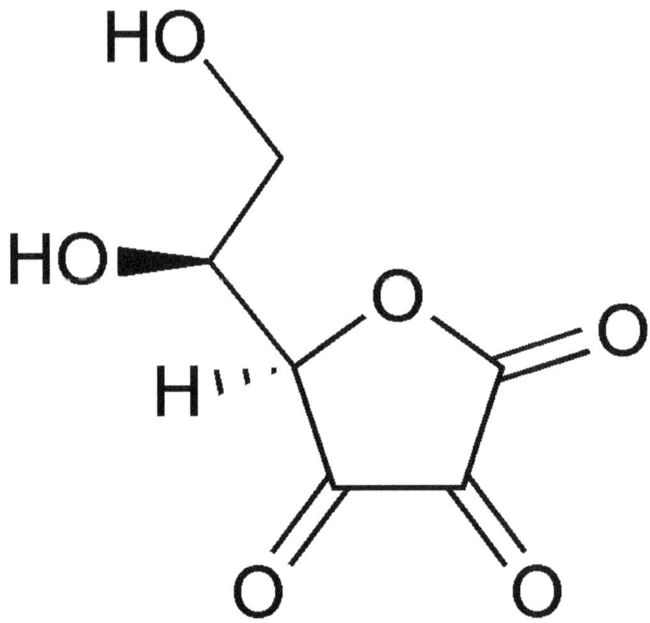

Top: ascorbic acid (reduced form of Vitamin C).

Bottom: dehydroascorbic acid (oxidized form of Vitamin C).

The process of cell respiration also depends heavily on the reduction of NAD^+ to NADH and the reverse reaction (the oxidation of NADH to NAD^+). Photosynthesis and cellular respiration are complementary, but photosynthesis is not the reverse of the redox reaction in cell respiration:

$$6\ CO_2 + 6\ H_2O + \text{light energy} \rightarrow C_6H_{12}O_6 + 6\ O_2$$

Biological energy is frequently stored and released by means of redox reactions. Photosynthesis involves the reduction of carbon dioxide into sugars and the oxidation of water into molecular oxygen. The reverse reaction, respiration, oxidizes sugars to produce carbon dioxide and water. As intermediate steps, the reduced carbon compounds are used to reduce nicotinamide adenine dinucleotide (NAD^+), which then contributes to the creation of a proton gradient, which drives the synthesis of adenosine triphosphate (ATP) and is maintained by the reduction of oxygen. In animal cells, mitochondria perform similar functions.

Free radical reactions are redox reactions that occur as a part of homeostasis and killing microorganisms, where an electron detaches from a molecule and then reattaches almost instantaneously. Free radicals are a part of redox molecules and can become harmful to the human body if they do not reattach to the redox molecule or an antioxidant. Unsatisfied free radicals can spur the mutation of cells they encounter and are, thus, causes of cancer.

The term redox state is often used to describe the balance of GSH/GSSG, NAD^+/NADH and $NADP^+$/NADPH in a biological system such as a cell or organ. The redox state is reflected in the balance of several sets of metabolites (e.g., lactate and pyruvate, beta-hydroxybutyrate, and acetoacetate), whose interconversion is dependent on these ratios. An abnormal redox state can develop in a variety of deleterious situations, such as hypoxia, shock, and sepsis. Redox mechanism also control some cellular processes. Redox proteins and their genes must be co-located for redox regulation according to the CoRR hypothesis for the function of DNA in mitochondria and chloroplasts.

Redox Cycling

A wide variety of aromatic compounds are enzymatically reduced to form free radicals that contain one more electron than their parent compounds. In general, the electron donor is any of a wide variety of flavoenzymes and their coenzymes. Once formed, these anion free radicals reduce molecular oxygen to superoxide, and regenerate the unchanged parent compound. The net reaction is the oxidation of the flavoenzyme's coenzymes and the reduction of molecular oxygen to form superoxide. This catalytic behavior has been described as futile cycle or redox cycling.

Examples of redox cycling-inducing molecules are the herbicide paraquat and other viologens and quinones such as menadione.

Redox Reactions in Geology

In geology, redox is important to both the formation of minerals and the mobilization of minerals, and is also important in some depositional environments.

Mi Vida uranium mine, near Moab, Utah. The alternating red and white/green bands of sandstone correspond to oxidized and reduced conditions in groundwater redox chemistry.

In general, the redox state of most rocks can be seen in the color of the rock. The rock forms in oxidizing conditions, giving it a red color. It is then "bleached" to a green—or sometimes white—form when a reducing fluid passes through the rock. The reduced fluid can also carry uranium-bearing minerals. Famous examples of redox conditions affecting geological processes include uranium deposits and Moqui marbles.

Balancing Redox Reactions

Describing the overall electrochemical reaction for a redox process requires a balancing of the component half-reactions for oxidation and reduction. In general, for reactions in aqueous solution, this involves adding H^+, OH^-, H_2O, and electrons to compensate for the oxidation changes.

Acidic Media

In acidic media, H^+ ions and water are added to half-reactions to balance the overall reaction.

For instance, when manganese(II) reacts with sodium bismuthate:

Unbalanced reaction:	$Mn^{2+}(aq) + NaBiO_3(s) \rightarrow Bi^{3+}(aq) +$ MnO– 4 (aq)
Oxidation:	$4\ H_2O(l) + Mn^{2+}(aq) \rightarrow$ MnO– 4(aq) $+ 8\ H^+(aq) + 5\ e^-$
Reduction:	$2\ e^- + 6\ H^+ +$ BiO– 3(s) $\rightarrow Bi^{3+}(aq) + 3\ H_2O(l)$

The reaction is balanced by scaling the two half-cell reactions to involve the same number of electrons (multiplying the oxidation reaction by the number of electrons in the reduction step and vice versa):

$8\ H_2O(l) + 2\ Mn^{2+}(aq) \rightarrow$ 2 MnO–4(aq) $+ 16\ H^+(aq) + 10\ e^-$

$10\ e^- + 30\ H^+ +$ 5 BiO–3(s) $\rightarrow 5\ Bi^{3+}(aq) + 15\ H_2O(l)$

Adding these two reactions eliminates the electrons terms and yields the balanced reaction:

$14\ H^+(aq) + 2\ Mn^{2+}(aq) + 5\ NaBiO_3(s) \rightarrow 7\ H_2O(l) +$ 2 MnO–4(aq) $+ 5\ Bi^{3+}(aq) +$ 5 Na+(aq)

Basic Media

In basic media, OH^- ions and water are added to half reactions to balance the overall reaction.

For example, in the reaction between potassium permanganate and sodium sulfite:

Unbalanced reaction:	$KMnO_4 + Na_2SO_3 + H_2O \rightarrow MnO_2 + Na_2SO_4 +$ KOH

Reduction:	$3\ e^- + 2\ H_2O + MnO-$ $4 \rightarrow MnO_2 + 4\ OH^-$
Oxidation:	$2\ OH^- + SO2-$ $3 \rightarrow SO2-$ $4 + H_2O + 2\ e^-$

Balancing the number of electrons in the two half-cell reactions gives:

$6\ e^- + 4\ H_2O + 2\ MnO{-}4 \rightarrow 2\ MnO_2 + 8\ OH^-$

$6\ OH^- + 3\ SO_2{-}3 \rightarrow 3\ SO_2{-}4 + 3\ H_2O + 6\ e^-$

Adding these two half-cell reactions together gives the balanced equation:

$2\ KMnO_4 + 3\ Na_2SO_3 + H_2O \rightarrow 2\ MnO_2 + 3\ Na_2SO_4 + 2\ KOH$

Memory Aids

The key terms involved in redox are often confusing to students. For example, an element that is oxidized loses electrons; however, that element is referred to as the reducing agent. Likewise, an element that is reduced gains electrons and is referred to as the oxidizing agent. Acronyms or mnemonics are commonly used to help remember the terminology:

- "OIL RIG" — **o**xidation **is** **l**oss of electrons, **r**eduction **is** **g**ain of electrons.
- "LEO the lion says GER" — **l**oss of **e**lectrons is **o**xidation, **g**ain of **e**lectrons is **r**eduction.
- "LEORA says GEROA" — **l**oss of **e**lectrons is **o**xidation (**r**educing **a**gent), **g**ain of **e**lectrons is **r**eduction (**o**xidizing **a**gent).
- "RED CAT" and "AN OX", or "AnOx RedCat" ("an ox-red cat") — **red**uction occurs at the **cat**hode and the **an**ode is for **ox**idation.
- "RED CAT gains what AN OX loses" – **red**uction at the **cat**hode gains (electrons) what **an**ode **ox**idation loses (electrons).

HYDROGENATION

Hydrogenation – to treat with hydrogen – is a chemical reaction between molecular hydrogen (H_2) and another compound or element, usually in the presence of a catalyst such as nickel, palladium or platinum. The process is commonly employed to reduce or saturate organic compounds. Hydrogenation typically constitutes the addition of pairs of hydrogenatoms to a molecule, often an alkene. Catalysts are required for the reaction to be usable; non-catalytic hydrogenation takes place only at very high temperatures. Hydrogenation reduces double and triple bonds in hydrocarbons.

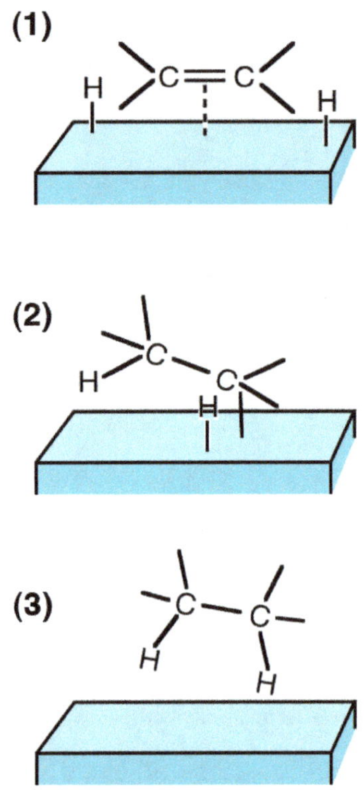

Steps in the hydrogenation of a C=C double bond at a catalyst surface, for example Ni or Pt :

(1) The reactants are adsorbed on the catalyst surface and H_2dissociates.
(2) An H atom bonds to one C atom. The other C atom is still attached to the surface.
(3) A second C atom bonds to an H atom. The molecule leaves the surface.

Process

Hydrogenation has three components, the unsaturated substrate, the hydrogen (or hydrogen source) and, invariably, a catalyst. The reduction reaction is carried out at different temperatures and pressures depending upon the substrate and the activity of the catalyst.

Related or Competing Reactions

The same catalysts and conditions that are used for hydrogenation reactions can also lead to isomerization of the alkenes from cis to trans. This process is of great interest because hydrogenation technology generates most of the trans fat in foods (see below). A reaction where bonds are broken while hydrogen is added is called hydrogenolysis, a reaction that may occur to carbon-carbon and carbon-heteroatom (oxygen, nitrogen or halogen) bonds. Some hydrogenations of polar bonds are accompanied by hydrogenolysis.

Hydrogen Sources

For hydrogenation, the obvious source of hydrogen is H_2 gas itself, which is typically available commercially within the storage medium of a pressurized cylinder. The hydrogenation process often uses greater than 1 atmosphere of H_2, usually conveyed from the cylinders and sometimes augmented by "booster pumps". Gaseous hydrogen is produced industrially from hydrocarbons by the process known as steam reforming. For many applications, hydrogen is transferred from donor molecules such as formic acid, isopropanol, and dihydroanthracene. These hydrogen donors undergo dehydrogenation to, respectively, carbon dioxide, acetone, and anthracene. These processes are called transfer hydrogenations.

Substrates

An important characteristic of alkene and alkyne hydrogenations, both the homogeneously and heterogeneously catalyzed versions, is that hydrogen addition occurs with "syn addition", with hydrogen entering from the least hindered side. Typical substrates are listed in the table

Substrates for and products of hydrogenation		
Substrate	**Product**	**Comments**
$R_2C{=}CR'_2$ (alkene)	$R_2CHCHR'_2$ (alkane)	many catalysts one application is margarine
RC≡CR (alkyne)	cis-RHC=CHR' (alkene)	over-hydrogenation to alkane can be problematic
RCHO (aldehyde)	RCH_2OH (primary alcohol)	easy substrate
R_2CO (ketone)	R_2CHOH (secondary alcohol)	more challenging than RCHO prochiral for unsymmetrical ketones
RCO_2R' (ester)	RCH_2OH + R'OH (two alcohols)	challenging substrate
RR'CNR" (imine)	RR'CHNHR" (amine)	easy substrate often use transfer hydrogenation actual precursor is N-protonated
$RC(O)NR'_2$ (amide)	$RCH_2NR'_2$ (amine)	challenging substrate
RCN (nitrile)	RCH_2NH_2 (primary amine)	product amine reactive toward precursor nitrile in some cases
RNO_2 (nitro)	RNH_2 (amine)	commercial applications use heterogeneous Ni and Ru catalysts major application is aniline

Catalysts

With rare exceptions, H_2 is unreactive toward organic compounds in the absence of metal catalysts. The unsaturated substrate is chemisorbed onto the catalyst, with most sites covered by the substrate. In heterogeneous catalysts, hydrogen forms surface hydrides (M-H) from which hydrogens can be transferred to the chemisorbed substrate. Platinum, palladium, rhodium, and ruthenium form highly active catalysts, which operate at lower temperatures and lower pressures of H_2. Non-precious metal catalysts, especially those based on nickel (such as Raney nickel and Urushibara nickel) have also been developed as economical alternatives, but they are often slower or require higher temperatures. The trade-off is activity (speed of reaction) vs. cost of the catalyst and cost of the apparatus required for use of high pressures. Notice that the Raney-nickel catalysed hydrogenations require high pressures:

Catalysts are usually classified into two broad classes: homogeneous catalysts and heterogeneous catalysts. Homogeneous catalysts dissolve in the solvent that contains the unsaturated substrate. Heterogeneous catalysts are solids that are suspended in the same solvent with the substrate or are treated with gaseous substrate.

Homogeneous Catalysts

Some well known homogeneous catalysts are indicated below. These are coordination complexes that activate both the unsaturated substrate and the H_2. Most typically, these complexes contain platinum group metals, especially Rh and Ir.

- Homogeneous hydrogenation catalysts and their precursors

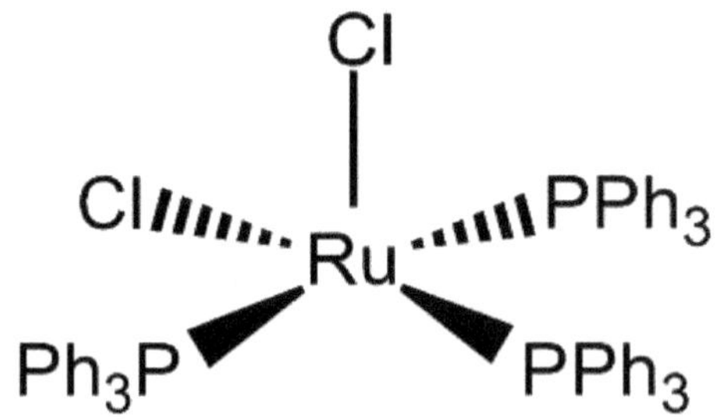

Dichlorotris(triphenylphosphine)ruthenium(II) is a precatalyst based on ruthenium.

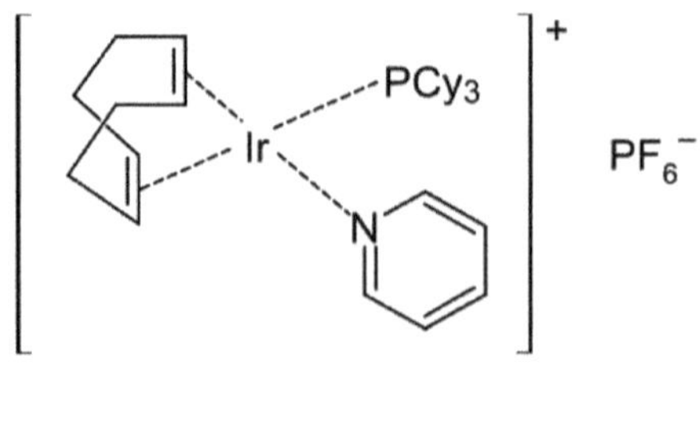

Crabtree's catalyst is a highly active catalyst featuring iridium.

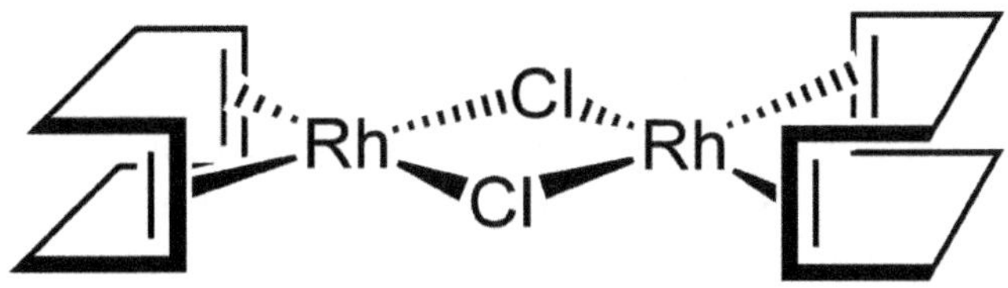

$Rh_2Cl_2(cod)_2$ is a precursor to many homogeneous catalysts.

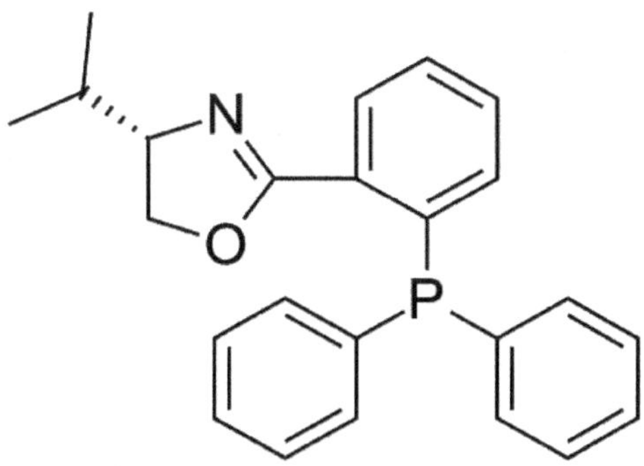

(S)-iPr-PHOX is a typical chelating phosphine ligand used in asymmetric hydrogenation.

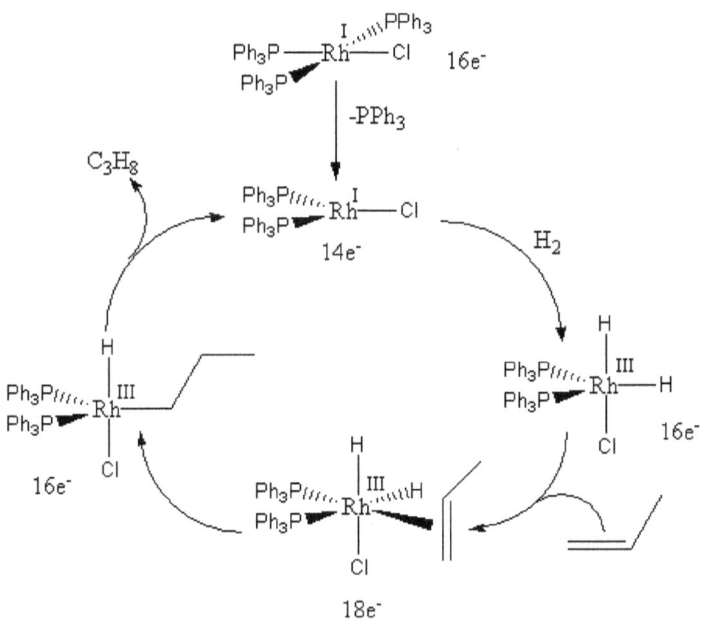

Catalytic hydrogenation of propylene

Hydrogenation of propylene with Wilkinson's catalyst.

Homogeneous catalysts are also used in asymmetric synthesis by the hydrogenation of prochiral substrates. An early demonstration of this approach was the Rh-catalyzed hydrogenation of enamides as precursors to the drug L-DOPA. To achieve asymmetric reduction, these catalyst are made chiral by use of chiral diphosphine ligands. Rhodium catalyzed hydrogenation has also been used in the herbicide production of S-metolachlor, which uses a Josiphos type ligand (called Xyliphos). In principle asymmetric hydrogenation can be catalyzed by chiral heterogeneous catalysts, but this approach remains more of a curiosity than a useful technology.

Heterogeneous Catalysts

Heterogeneous catalysts for hydrogenation are more common industrially. As in homogeneous catalysts, the activity is adjusted through changes in the environment around the metal, i.e. the coordination sphere. Different faces of a crystalline heterogeneous catalyst display distinct activities, for example. Similarly, heterogeneous catalysts are affected by their supports, i.e. the material upon with the heterogeneous catalyst is bound.

In many cases, highly empirical modifications involve selective "poisons". Thus, a carefully chosen catalyst can be used to hydrogenate some functional groups without affecting others, such as the hydrogenation of alkenes without touching aromatic rings, or the selective hydrogenation of alkynes to alkenes using Lindlar's catalyst. For example, when the catalyst palladium is placed on barium sulfate and then treated with quinoline, the resulting catalyst reduces alkynes only as far as alkenes. The Lindlar catalyst has been applied to the conversion of phenylacetylene to styrene.

- Illustrative hydrogenations

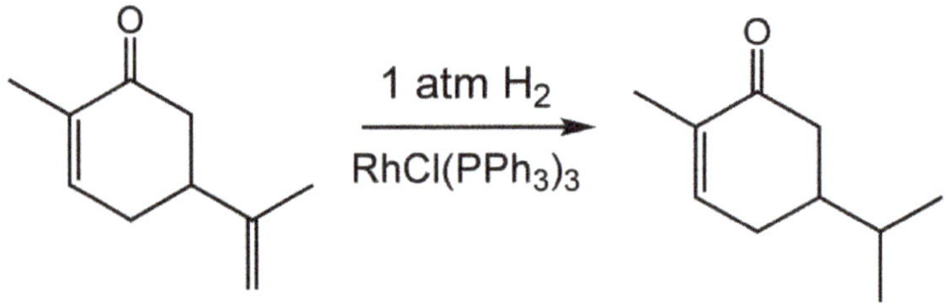

Selective hydrogenation of the less hindered alkene group in carvone using a homogeneous catalyst (Wilkinson's catalyst).

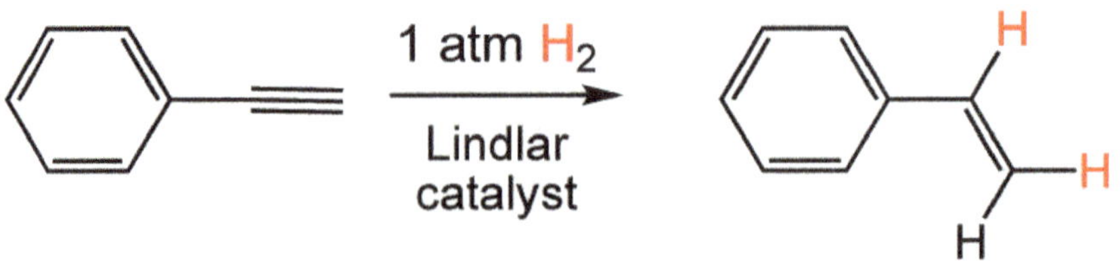

Partial hydrogenation of phenylacetylene using the Lindlar catalyst.

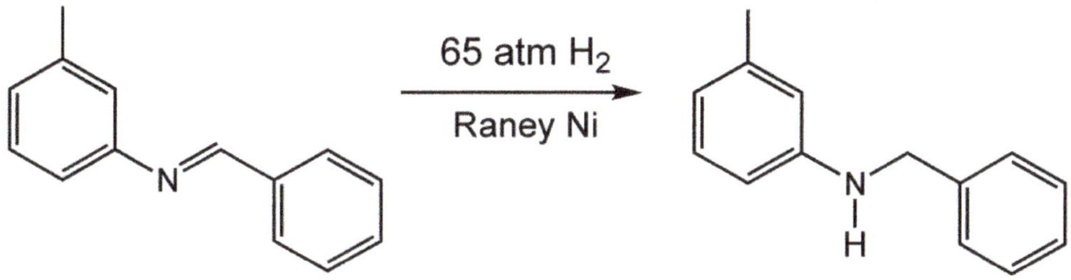

Hydrogenation of an imine using a Raney nickel catalyst, a popular heterogeneous catalyst.

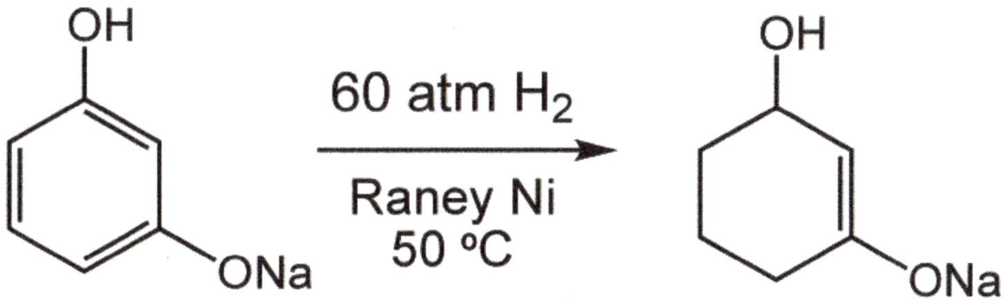

Partial hydrogenation of a resorcinol derivative using a Raney-Nickel catalyst.

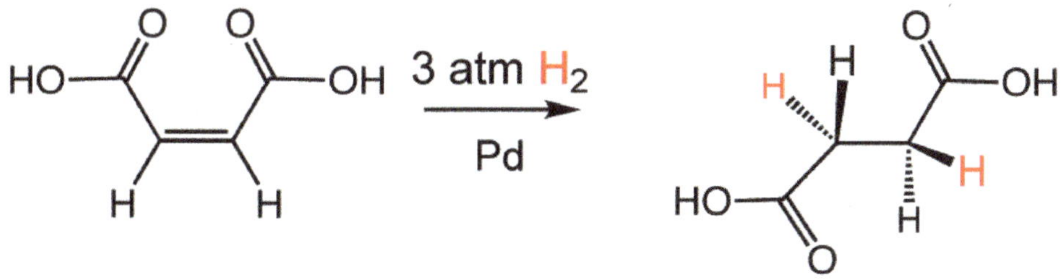

Hydrogenation of maleic acid to succinic acid.

Transfer Hydrogenation

Hydrogen also can be extracted ("transferred") from "hydrogen-donors" in place of H_2 gas. Hydrogen donors, which often serve as solvents include hydrazine, dihydronaphthalene, dihydroanthracene, isopropanol, and formic acid.

In organic synthesis, transfer hydrogenation is useful for the asymmetric reduction of polar unsaturated substrates, such as ketones, aldehydes, and imines. The hydrogenation of polar substrates such as ketones and aldehydes typically require transfer hydrogenation, at least in homogeneous catalysis. These catalysts are readily generated in chiral forms, which is the basis of asymmetric hydrogenation of ketones.

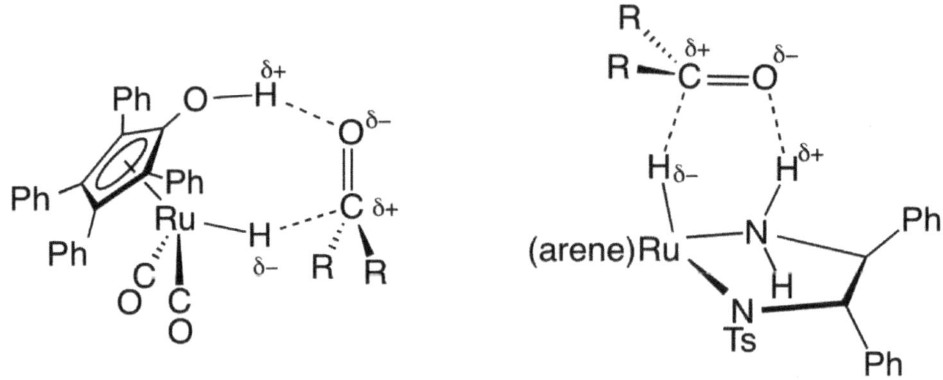

Transfer hydrogenation catalyzed by transition metal complexes proceeds by an "outer sphere mechanism."

Electrolytic Hydrogenation

Polar substrates such as nitriles can be hydrogenated electrochemically, using protic solvents and reducing equivalents as the source of hydrogen.

Thermodynamics and Mechanism

The addition of hydrogen to double or triple bonds in hydrocarbons is a type of redox reaction that can be thermodynamically favorable. For example, the addition of hydrogen to an alkene has a Gibbs free energy change of -101 kJ·mol^{-1}. However, the reaction rate for most hydrogenation reactions is negligible in the absence of catalysts. Hydrogenation is a strongly exothermic reaction. In the hydrogenation of vegetable oils and fatty acids, for example, the heat released is about 25 kcal per mole (105 kJ/mol), sufficient to raise the temperature of the oil by 1.6–1.7 °C per iodine number drop. The mechanism of metal-catalyzed hydrogenation of alkenes and alkynes has been extensively studied. First of all isotope labeling using deuterium confirms the regiochemistry of the addition:

$RCH{=}CH_2 + D_2 \rightarrow RCHDCH_2D$

Heterogeneous Catalysis

On solids, the accepted mechanism is the Horiuti-Polanyi mechanism:

- Binding of the unsaturated bond, and hydrogen dissociation into atomic hydrogen onto the catalyst
- Addition of one atom of hydrogen; this step is reversible
- Addition of the second atom; effectively irreversible under hydrogenating conditions.

In the second step, the metallointermediate formed is a saturated compound that can rotate and then break down, again detaching the alkene from the catalyst. Consequently, contact with a hydrogenation catalyst necessarily causes cis-trans-isomerization, because the isomerization is thermodynamically favorable. This is a problem in partial hydrogenation, while in complete hydrogenation the produced trans-alkene is eventually hydrogenated.

For aromatic substrates, the first bond is hardest to hydrogenate because of the free energy penalty for breaking the aromatic system. The product of this is a cyclohexadiene, which is extremely active and cannot be isolated; in conditions reducing enough to break the aromatization, it is immediately reduced to a cyclohexene. The cyclohexene is ordinarily reduced immediately to a fully saturated cyclohexane, but special modifications to the catalysts (such as the use of the anti-solvent water on ruthenium) can preserve some of the cyclohexene, if that is a desired product.

Homogeneous Catalysis

In many homogeneous hydrogenation processes, the metal binds to both components to give an intermediate alkene-metal$(H)_2$ complex. The general sequence of reactions is assumed to be as follows or a related sequence of steps:

- binding of the hydrogen to give a dihydride complex via oxidative addition (preceding the oxidative addition of H_2 is the formation of a dihydrogen complex):

$L_nM + H_2 \rightarrow L_nMH_2$

- binding of alkene:

$L_nM(\eta^2H_2) + CH_2{=}CHR \rightarrow L_{n-1}MH_2(CH_2{=}CHR) + L$

- transfer of one hydrogen atom from the metal to carbon (migratory insertion)

$L_{n-1}MH_2(CH_2{=}CHR) \rightarrow L_{n-1}M(H)(CH_2\text{-}CH_2R)$

- transfer of the second hydrogen atom from the metal to the alkyl group with simultaneous dissociation of the alkane ("reductive elimination")

$L_{n-1}M(H)(CH_2\text{-}CH_2R) \rightarrow L_{n-1}M + CH_3\text{-}CH_2R$

Inorganic Substrates

The hydrogenation of nitrogen to give ammonia is conducted on a vast scale by the Haber-Bosch process, consuming an estimated 1% of the world's energy supply.

$$\overbrace{N{\equiv}N}^{\text{nitrogen}} + \overbrace{3\,H_2}^{\substack{\text{hydrogen}\\(200\ \text{atm})}} \xrightarrow[350-550^\circ C]{\text{Fe catalyst}} \overbrace{2\,NH_3}^{\text{ammonia}}$$

Oxygen can be partially hydrogenated to give hydrogen peroxide, although this process has not been commercialized.

Industrial Applications

Catalytic hydrogenation has diverse industrial uses. Most frequently, industrial hydrogenation relies on heterogeneous catalysts.

Food Industry

Types of fats in food

• Unsaturated fat
  • Monounsaturated fat
  • Polyunsaturated fat
  • Trans fat
  • Omega numbering:
    • ω−3
    • ω−6
    • ω−7
    • ω−9
• Saturated fat
  • Interesterified fat

The largest scale application of hydrogenation is for the processing of vegetable oils. Typical vegetable oils are derived from polyunsaturated fatty acids (containing more than one carbon-carbon double bond). Their partial hydrogenation reduces most but not all, of these carbon-carbon double bonds. The degree of hydrogenation is controlled by restricting the amount of hydrogen, reaction temperature and time, and the catalyst.

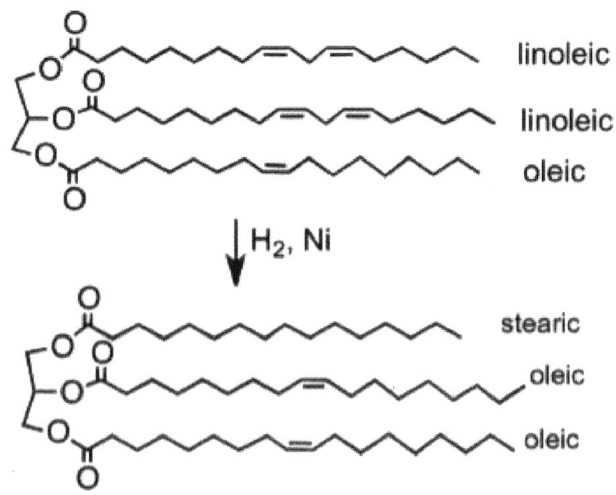

Partial hydrogenation of a typical plant oil to a typical component of margarine. Most of the C=C double bonds are removed in this process, which elevates the melting point of the product.

Hydrogenation converts liquid vegetable oils into solid or semi-solid fats, such as those present in margarine. Changing the degree of saturation of the fat changes some important physical properties, such as the melting range, which is why liquid oils become semi-solid. Solid or semi-solid fats are preferred for baking because the way the fat mixes with flour produces a more desirable texture in the baked product. Because partially hydrogenated vegetable oils are cheaper than animal fats, are available in a wide range of consistencies, and have other desirable characteristics (such as increased oxidative stability and longer shelf life), they are the predominant fats used as shortening in most commercial baked goods.

A side effect of incomplete hydrogenation having implications for human health is the isomerization of some of the remaining unsaturated carbon bonds, resulting in the trans isomers, which have been implicated in circulatory diseases including heart disease. The conversion from cis to trans bonds is favored because the trans configuration has lower energy than the natural cis one. At equilibrium, the trans/cis isomer ratio is about 2:1. Many countries and regions have introduced mandatory labeling of trans fats on food products and appealed to the industry for voluntary reductions.

Petrochemical Industry

In petrochemical processes, hydrogenation is used to convert alkenes and aromatics into saturated alkanes (paraffins) and cycloalkanes (naphthenes), which are less toxic and less reactive. Relevant to liquid fuels that are stored sometimes for long periods in air, saturated hydrocarbons exhibit superior storage properties. On the other hand, alkene tend to form hydroperoxides, which can form gums that interfere with fuel handing equipment. For example, mineral turpentine is usually hydrogenated. Hydrocracking of heavy residues into diesel is another application. In isomerization and catalytic reforming processes, some hydrogen pressure is maintained to hydrogenolyze coke formed on the catalyst and prevent its accumulation.

Organic Chemistry

Hydrogenation is a useful means for converting unsaturated compounds into saturated derivatives. Substrates include not only alkenes and alkynes, but also aldehydes, imines, and nitriles, which are converted into the corresponding saturated compounds, i.e. alcohols and amines. Thus, alkyl aldehydes, which can be synthesized with the oxo process from carbon monoxide and an alkene, can be converted to alcohols. E.g. 1-propanol is produced from propionaldehyde, produced from ethene and carbon monoxide. Xylitol, a polyol, is produced by hydrogenation of the sugar xylose, an aldehyde. Primary amines can be synthesized by hydrogenation of nitriles, while nitriles are readily synthesized from cyanide and a suitable electrophile. For example, isophorone diamine, a precursor to the polyurethane monomer isophorone diisocyanate, is produced from isophorone nitrile by a tandem nitrile hydrogenation/reductive amination by ammonia, wherein hydrogenation converts both the nitrile into an amine and the imine formed from the aldehyde and ammonia into another amine.

Hydrogenation of Coal

Heterogeneous Catalytic Hydrogenation

The earliest hydrogenation is that of platinum catalyzed addition of hydrogen to oxygen in the Döbereiner's lamp, a device commercialized as early as 1823. The French chemist Paul Sabatier is considered the father of the hydrogenation process. In 1897, building on the earlier work of James Boyce, an American chemist working in the manufacture of soap products, he discovered that traces of nickel catalyzed the addition of hydrogen to molecules of gaseous hydrocarbons in what is now known as the Sabatier process. For this work, Sabatier shared the 1912 Nobel Prize in Chemistry. Wilhelm Normann was awarded a patent in Germany in 1902 and in Britain in 1903 for the hydrogenation of liquid oils, which was the beginning of what is now a worldwide industry. The commercially important Haber–Bosch process, first described in 1905, involves hydrogenation of nitrogen. In the Fischer–Tropsch process, reported in 1922 carbon monoxide, which is easily derived from coal, is hydrogenated to liquid fuels.

In 1922, Voorhees and Adams described an apparatus for performing hydrogenation under pressures above one atmosphere. The Parr shaker, the first product to allow hydrogenation using elevated pressures and temperatures, was commercialized in 1926 based on Voorhees and Adams' research and remains in widespread use. In 1924 Murray Raney developed a finely powdered form of nickel, which is widely used to catalyze hydrogenation reactions such as conversion of nitriles to amines or the production of margarine.

Homogeneous Catalytic Hydrogenation

In the 1930s, Calvin discovered that copper(II) complexes oxidized H_2. The 1960s witnessed the development of well defined homogeneous catalysts using transition metal complexes, e.g., Wilkinson's catalyst ($RhCl(PPh_3)_3$). Soon thereafter cationic Rh and Ir were found catalyze the hydrogenation of alkenes and carbonyls. In the 1970s, asymmetric hydrogenation was demonstrated in the synthesis of L-DOPA, and the 1990s saw the invention of Noyori asymmetric hydrogenation. The development of homogeneous hydrogenation was influenced by work started in the 1930s and 1940s on the oxo process and Ziegler–Natta polymerization.

Metal-Free Hydrogenation

For most practical purposes, hydrogenation requires a metal catalyst. Hydrogenation can, however, proceed from some hydrogen donors without catalysts, illustrative hydrogen donors being diimide and aluminium isopropoxide, the latter illustrated by the Meerwein–Ponndorf–Verley reduction. Some metal-free catalytic systems have been investigated in academic research. One such system for reduction of ketones consists of tert-butanol and potassium tert-butoxide and very high temperatures. The reaction depicted below describes the hydrogenation of benzophenone:

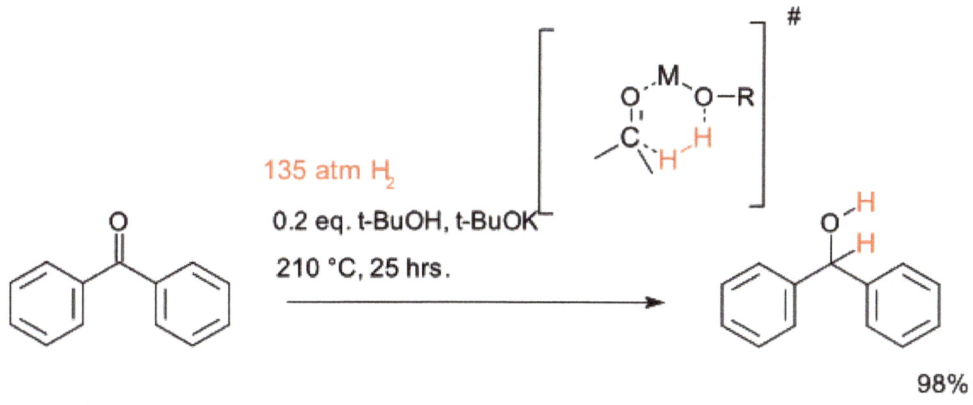

A chemical kinetics study found this reaction is first-order in all three reactants suggesting a cyclic 6-membered transition state.

Another system for metal-free hydrogenation is based on the phosphine-borane, compound **1**, which has been called a frustrated Lewis pair. It reversibly accepts dihydrogen at relatively low temperatures to form the phosphonium borate **2** which can reduce simple hindered imines.

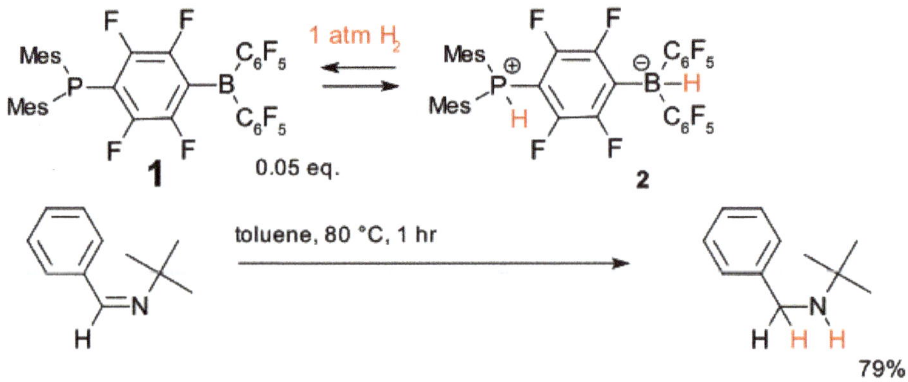

The reduction of nitrobenzene to aniline has been reported to be catalysed by fullerene, its mono-anion, atmospheric hydrogen and UV light.

Equipment Used for Hydrogenation

Today's bench chemist has three main choices of hydrogenation equipment:

- Batch hydrogenation under atmospheric conditions
- Batch hydrogenation at elevated temperature and/or pressure
- Flow hydrogenation

Batch Hydrogenation Under Atmospheric Conditions

The original and still a commonly practiced form of hydrogenation in teaching laboratories, this process is usually effected by adding solid catalyst to a round bottom flask of dissolved reactant which has been evacuated using nitrogen or argon gas and sealing the mixture with a penetrable rubber seal. Hydrogen gas is then supplied from a H_2-filled balloon. The resulting three phase mixture is agitated to promote mixing. Hydrogen uptake can be monitored, which can be useful for monitoring progress of a hydrogenation. This is achieved by either using a graduated tube containing a coloured liquid, usually aqueous copper sulfate or with gauges for each reaction vessel.

Batch Hydrogenation At Elevated Temperature and/or Pressure

Since many hydrogenation reactions such as hydrogenolysis of protecting groups and the reduction of aromatic systems proceed extremely sluggishly at atmospheric temperature and pressure, pressurized systems are popular. In these cases, catalyst is added to a solution of reactant under an inert atmosphere in a pressure vessel. Hydrogen is added directly from a cylinder or built in laboratory hydrogen source, and the pressurized slurry is mechanically rocked to provide agitation, or a spinning basket is used. Heat may also be used, as the pressure compensates for the associated reduction in gas solubility.

Flow Hydrogenation

Flow hydrogenation has become a popular technique at the bench and increasingly the process scale. This technique involves continuously flowing a dilute stream of dissolved reactant over a fixed bed catalyst in the presence of hydrogen. Using established HPLC technology, this technique allows the application of pressures from atmospheric to 1,450 psi (100 bar). Elevated temperatures may also be used. At the bench scale, systems use a range of pre-packed catalysts which eliminates the need for weighing and filtering pyrophoric catalysts.

Industrial Reactors

Catalytic hydrogenation is done in a tubular plug-flow reactor (PFR) packed with a supported catalyst. The pressures and temperatures are typically high, although this depends on the catalyst. Catalyst loading is typically much lower than in laboratory batch hydrogenation, and various promoters are added to the metal, or mixed metals are used, to improve activity, selectivity and catalyst stability. The use of nickel is common despite its low activity, due to its low cost compared to precious metals.

Gas Liquid Induction Reactors (Hydrogenator) are also used for carrying out catalytic hydrogenation.

DEHYDROGENATION

Dehydrogenation is a chemical reaction that involves the removal of hydrogen from an organic molecule. It is the reverse of hydrogenation. Dehydrogenation is an important reaction because it converts alkanes, which are relatively inert and thus low-valued, to olefins, which are reactive and thus more valuable. Alkenes are precursors to aldehydes, alcohols, polymers, and aromatics. Dehydrogenation processes are used extensively to produce aromatics and styrene in the petrochemical industry. Such processes are highly endothermic and require temperatures of 500 °C and above. Dehydrogenation also converts saturated fats to unsaturated fats. Enzymes that catalyze dehydrogenation are called dehydrogenases.

Classes of the Reaction

A variety of dehydrogenation processes have been described, especially for organic compounds:

- In typical aromatization, six-membered alicyclic rings, e.g. cyclohexene, can be aromatized in the presence of hydrogenation acceptors. The elements sulfur and selenium promote this process. On the laboratory scale, quinones, especially 2,3-Dichloro-5,6-dicyano-1,4-benzoquinone are effective.
- Oxidation of alcohols to ketones or aldehydes can be effected by metal catalysts such as copper chromite. In the Oppenauer oxidation, hydrogen is transferred from an alcohol to an aldehyde or ketone to bring about the oxidation.

- Dehydrogenation of amines to nitriles using a variety of reagents, such as Iodine pentafluoride (IF5).
- Dehydrogenation of paraffins and olefins — paraffins such as n-pentane and isopentane can be converted to pentene and isopentene using chromium (III) oxide as a catalyst at 500 °C.

Examples

One of the largest scale dehydrogenation reactions is the production of styrene by dehydrogenation of ethylbenzene. Typical dehydrogenation catalysts are based on iron(III) oxide, promoted by several percent potassium oxide or potassium carbonate.

$$C_6H_5CH_2CH_3 \rightarrow C_6H_5CH{=}CH_2 + H_2$$

Formaldehyde is produced industrially by the catalytic oxidation of methanol, which can also be viewed as a dehydrogenation using O_2 as the acceptor. The most common catalysts are silver metal or a mixture of an iron and molybdenum or vanadium oxides. In the commonly used formox process, methanol and oxygen react at ca. 250–400 °C in presence of iron oxide in combination with molybdenum and/or vanadium to produce formaldehyde according to the chemical equation:

$$2\ CH_3OH + O_2 \rightarrow 2\ CH_2O + 2\ H_2O$$

The importance of catalytic dehydrogenation of paraffin hydrocarbons to olefins has been growing steadily in recent years. Light olefins, such as butenes, are important raw materials for the synthesis of polymers, gasoline additives and various other petrochemical products. The cracking processes especially fluid catalytic cracking and steam cracker produce high-purity mono-olefins, such as 1-butene or butadiene. Despite such processes, currently more research is focused on developing alternatives such as oxidative dehydrogenation (ODH) for two reasons: (1) undesired reactions take place at high temperature leading to coking and catalyst deactivation, making frequent regeneration of the catalyst unavoidable, (2) it consumes a large amount of heat and requires high reaction temperatures. Oxidative dehydrogenation (ODH) of n-butane is an alternative to classical dehydrogenation, steam cracking and fluid catalytic cracking processes.

- Butane
- Butene
- Cyclohexane
- Propane

Homogeneous Catalysis

Although not of commercial value, dehydrogenation of alkanes can be effected by homogeneous catalysis. Especially active for this reaction are pincer complexes.

HYDROLYSIS

Hydrolysis usually means the cleavage of chemical bonds by the addition of water. When a carbohydrate is broken into its component sugar molecules by hydrolysis (e.g. sucrose being broken down into glucose and fructose), this is termed saccharification. Generally, hydrolysis or saccharification is a step in the degradation of a substance.

Hydrolysis can be the reverse of a condensation reaction in which two molecules join together into a larger one and eject a water molecule. Thus hydrolysis adds water to break down, whereas condensation builds up by removing water.

Generic mechanism for a hydrolysis reaction. (The 2-way harpoons as the yield symbol indicates an equilibrium in which hydrolysis and condensation can go both ways.).

Types

Usually hydrolysis is a chemical process in which a molecule of water is added to a substance. Sometimes this addition causes both substance and water molecule to split into two parts. In such reactions, one fragment of the target molecule (or parent molecule) gains a hydrogen ion.

Salts

A common kind of hydrolysis occurs when a salt of a weak acid or weak base (or both) is dissolved in water. Water spontaneously ionizes into hydroxide anions and hydronium cations. The salt also dissociates into its constituent anions and cations. For example, sodium acetate dissociates in water into sodium and acetate ions. Sodium ions react very little with the hydroxide ions whereas the acetate ions combine with hydronium ions to produce acetic acid. In this case the net result is a relative excess of hydroxide ions, yielding a basic solution.

Strong acids also undergo hydrolysis. For example, dissolving sulfuric acid (H_2SO_4) in water is accompanied by hydrolysis to give hydronium and bisulfate, the sulfuric acid's conjugate base. For a more technical discussion of what occurs during such a hydrolysis.

Esters and Amides

Acid–base-catalysed hydrolyses are very common; one example is the hydrolysis of amides or esters. Their hydrolysis occurs when the nucleophile (a nucleus-seeking agent, e.g., water or hydroxyl ion) attacks the carbon of the carbonyl group of the ester or amide.

In an aqueous base, hydroxyl ions are better nucleophiles than polar molecules such as water. In acids, the carbonyl group becomes protonated, and this leads to a much easier nucleophilic attack. The products for both hydrolyses are compounds with carboxylic acid groups.

Perhaps the oldest commercially practiced example of ester hydrolysis is saponification (formation of soap). It is the hydrolysis of a triglyceride (fat) with an aqueous base such as sodium hydroxide (NaOH). During the process, glycerol is formed, and the fatty acids react with the base, converting them to salts. These salts are called soaps, commonly used in households.

In addition, in living systems, most biochemical reactions (including ATP hydrolysis) take place during the catalysis of enzymes. The catalytic action of enzymes allows the hydrolysis of proteins, fats, oils, and carbohydrates. As an example, one may consider proteases (enzymes that aid digestion by causing hydrolysis of peptide bonds in proteins). They catalyze the hydrolysis of interior peptide bonds in peptide chains, as opposed to exopeptidases (another class of enzymes, that catalyze the hydrolysis of terminal peptide bonds, liberating one free amino acid at a time).

However, proteases do not catalyze the hydrolysis of all kinds of proteins. Their action is stereo-selective: Only proteins with a certain tertiary structure are targeted as some kind of orienting force is needed to place the amide group in the proper position for catalysis. The necessary contacts between an enzyme and its substrates (proteins) are created because the enzyme folds in such a way as to form a crevice into which the substrate fits; the crevice also contains the catalytic groups. Therefore, proteins that do not fit into the crevice will not undergo hydrolysis. This specificity preserves the integrity of other proteins such as hormones, and therefore the biological system continues to function normally.

Upon hydrolysis, an amide converts into a carboxylic acid and an amine or ammonia (which in the presence of acid are immediately converted to ammonium salts). One of the two oxygen groups on the carboxylic acid are derived from a water molecule and the amine (or ammonia) gains the hydrogen ion. The hydrolysis of peptides gives amino acids.

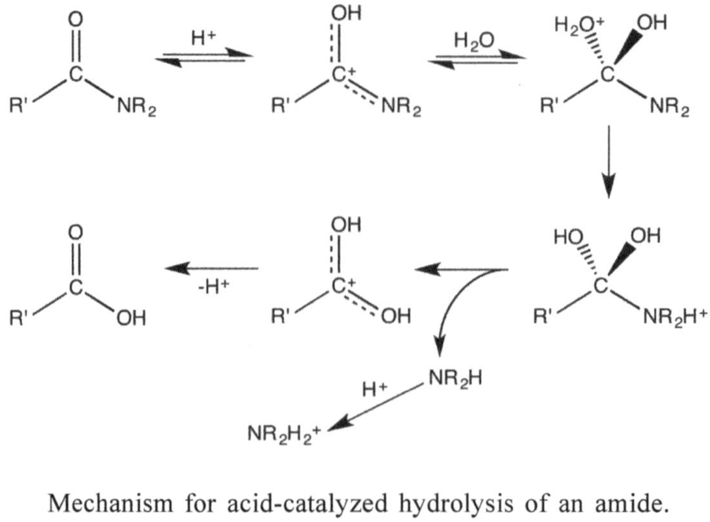

Mechanism for acid-catalyzed hydrolysis of an amide.

Many polyamide polymers such as nylon 6,6 hydrolyze in the presence of strong acids. The process leads to depolymerization. For this reason nylon products fail by fracturing when exposed to small amounts of acidic water. Polyesters are also susceptible to similar polymer degradation reactions. The problem is known as environmental stress cracking.

ATP

Hydrolysis is related to energy metabolism and storage. All living cells require a continual supply of energy for two main purposes: the biosynthesis of micro and macromolecules, and the active transport of ions and molecules across cell membranes. The energy derived from the oxidation of nutrients is not used directly but, by means of a complex and long sequence of reactions, it is channelled into a special energy-storage molecule, adenosine triphosphate (ATP). The ATP molecule contains pyrophosphate linkages (bonds formed when two phosphate units are combined together) that release energy when needed. ATP can undergo hydrolysis in two ways: the removal of terminal phosphate to form adenosine diphosphate (ADP) and inorganic phosphate, or the removal of a terminal diphosphate to yield adenosine monophosphate (AMP) and pyrophosphate. The latter usually undergoes further cleavage into its two constituent phosphates. This results in biosynthesis reactions, which usually occur in chains, that can be driven in the direction of synthesis when the phosphate bonds have undergone hydrolysis.

Polysaccharides

Monosaccharides can be linked together by glycosidic bonds, which can be cleaved by hydrolysis. Two, three, several or many monosaccharides thus linked form disaccharides, trisaccharides, oligosaccharides or polysaccharides, respectively. Enzymes that hydrolyse glycosidic bonds are called "glycoside hydrolases" or "glycosidases".

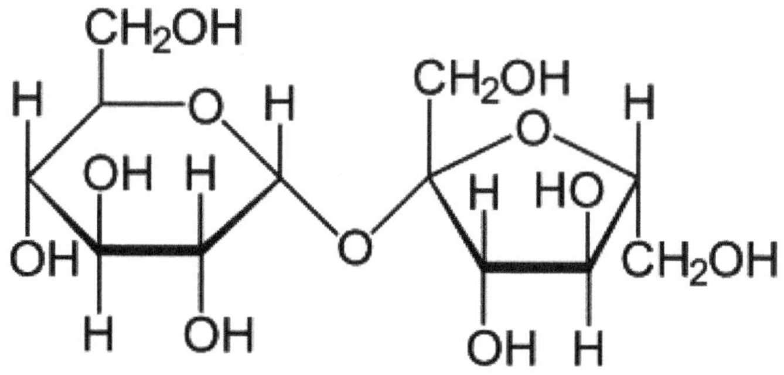

Sucrose. The glycoside bond is represented by the central oxygen atom, which holds the two monosaccharide units together.

The best-known disaccharide is sucrose (table sugar). Hydrolysis of sucrose yields glucose and fructose. Invertase is a sucrase used industrially for the hydrolysis of sucrose to so-called invert sugar. Lactase is essential for digestive hydrolysis of lactose in milk;

many adult humans do not produce lactase and cannot digest the lactose in milk (not a disorder).

The hydrolysis of polysaccharides to soluble sugars is called "saccharification". Malt made from barley is used as a source of β-amylase to break down starch into the disaccharide maltose, which can be used by yeast to produce beer. Other amylase enzymes may convert starch to glucose or to oligosaccharides. Cellulose is first hydrolyzed to cellobiose by cellulase and then cellobiose is further hydrolyzed to glucose by beta-glucosidase. Animals such as cows (ruminants) are able to hydrolyze cellulose into cellobiose and then glucose because of symbiotic bacteria that produce cellulases.

Metal Aqua Ions

Metal ions are Lewis acids, and in aqueous solution they form metal aqua ions of the general formula $M(H_2O)_n^{m+}$. The aqua ions undergo hydrolysis, to a greater or lesser extent. The first hydrolysis step is given generically as

$$M(H_2O)_n^{m+} + H_2O \rightleftharpoons M(H_2O)_{n-1}(OH)^{(m-1)+} + H_3O^+$$

Thus the aqua cations behave as acids in terms of Brønsted-Lowry acid-base theory. This effect is easily explained by considering the inductive effect of the positively charged metal ion, which weakens the O-H bond of an attached water molecule, making the liberation of a proton relatively easy.

The dissociation constant, pK_a, for this reaction is more or less linearly related to the charge-to-size ratio of the metal ion. Ions with low charges, such as Na^+ are very weak acids with almost imperceptible hydrolysis. Large divalent ions such as Ca^{2+}, Zn^{2+}, Sn^{2+} and Pb^{2+} have a pK_a of 6 or more and would not normally be classed as acids, but small divalent ions such as Be^{2+} undergo extensive hydrolysis. Trivalent ions like Al^{3+} and Fe^{3+} are weak acids whose pK_a is comparable to that of acetic acid. Solutions of salts such as $BeCl_2$ or $Al(NO_3)_3$ in water are noticeably acidic; the hydrolysis can be suppressed by adding an acid such as nitric acid, making the solution more acidic.

Hydrolysis may proceed beyond the first step, often with the formation of polynuclear species via the process of olation. Some "exotic" species such as $Sn_3(OH)_4^{2+}$ are well characterized. Hydrolysis tends to proceed as pH rises leading, in many cases, to the precipitation of a hydroxide such as $Al(OH)_3$ or $AlO(OH)$. These substances, major constituents of bauxite, are known as laterites and are formed by leaching from rocks of most of the ions other than aluminium and iron and subsequent hydrolysis of the remaining aluminium and iron.

Hydration Reaction

In chemistry, a hydration reaction is a chemical reaction in which a substance combines with water. In organic chemistry, water is added to an unsaturated substrate, which is usually an alkene or an alkyne. This type of reaction is employed industrially to produce ethanol, isopropanol, and 2-butanol.

Epoxides to Glycol

Several billion kilograms of ethylene glycol is produced annually by the hydration of ethylene oxide:

$C_2H_4O + H_2O \rightarrow HO–CH_2CH_2–OH$

Acid catalysts are typically used.

Alkenes

For the hydration of alkenes, the general chemical equation of the reaction is the following:

$RRCH=CH_2$ in H_2O/acid $\rightarrow$ RRCH(-OH)-CH_3

A hydroxyl group (OH^-) attaches to one carbon of the double bond, and a proton (H^+) adds to the other carbon of the double bond. The reaction is highly exothermic. In the first step, the alkene acts as a nucleophile and attacks the proton, following Markovnikov's rule. In the second step an H_2O molecule bonds to the other, more highly substituted carbon. The oxygen atom at this point has three bonds and carries a positive charge (i.e., the molecule is an oxonium). Another water molecule comes along and takes up the extra proton. This reaction tends to yield many undesirable side products, (for example diethyl ether in the process of creating Ethanol) and in its simple form described here is not considered very useful for the production of alcohol.

Two approaches are taken. Traditionally the alkene is treated with sulfuric acid to give alkyl sulfate esters. In the case of ethanol production, this step can be written:

$H_2SO_4 + C_2H_4 \rightarrow C_2H_5\text{-O-}SO_3H$

Subsequently this sulfate ester is hydrolyzed to regenerate sulfuric acid and release ethanol:

$C_2H_5\text{-O-}SO_3H + H_2O \rightarrow H_2SO_4 + C_2H_5OH$

This two step route is called the "indirect process".

In the "direct process," the acid protonates the alkene, and water reacts with this incipient carbocation to give the alcohol. The direct process is more popular because it is simpler. The acid catalysts include phosphoric acid and several solid acids. Here an example reaction mechanism of the hydration of 1-methylcyclohexene to 1-methylcyclohexanol:

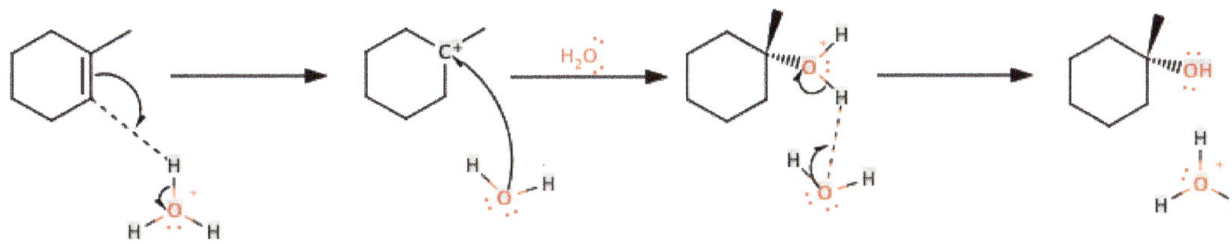

Many alternative routes are available for producing alcohols, including the hydroboration–oxidation reaction, the oxymercuration–reduction reaction, fermentation and reduction of ketones and aldehydes.

Hydration of other Substrates

Any unsaturated organic compound is susceptible to hydration. Acetylene hydrates to give acetaldehyde: The process typically relies on mercury catalysts and has been discontinued in the West but is still practiced in China. The Hg^{2+} center binds to C≡C triple bond, which is then attacked by water. The reaction is:

$H_2O + C_2H_2 \rightarrow CH_3CHO$

Nitriles undergo hydration to give amides:

$H_2O + RCN \rightarrow RC(O)NH_2$

This reaction is employed in the production of acrylamide.

Aldehydes and to some extent even ketones, hydrate to geminal diols. The reaction is especially dominant for formaldehyde, which, in the presence of water, exists significantly as dihydroxymethane.

Conceptually similar reactions include hydroamination and hydroalkoxylation, which involve adding amines and alcohols to alkenes.

Inorganic and Materials Chemistry

Hydration is an important process in many other applications; one example is the production of Portland cement by the crosslinking of calcium oxides and silicates that is induced by water. Hydration is the process by which desiccants function.

Dehydration Reaction

In chemistry and the biological sciences, a dehydration reaction is usually defined as a chemical reaction that involves the loss of a water molecule from the reacting molecule. Dehydration reactions are a subset of condensation reactions. Because the hydroxyl group (–OH) is a poor leaving group, having a Brønsted acid catalyst often helps by protonating the hydroxyl group to give the better leaving group, $-OH_2^+$. The reverse of a dehydration reaction is a hydration reaction. Common dehydrating agents used in organic synthesis include concentrated sulfuric acid, concentrated phosphoric acid, hot aluminium oxide and hot ceramic.

Dehydration reactions and dehydration synthesis have the same meaning, and are often used interchangeably. Two monosaccharides, such as glucose and fructose, can be joined together (to form sucrose) using dehydration synthesis. The new molecule, consisting of two monosaccharides, is called a disaccharide.

The process of hydrolysis is the reverse reaction, meaning that the water is recombined with the two hydroxyl groups and the disaccharide reverts to being monosaccharides.

In the related condensation reaction water is released from two different reactants.

Dehydration Reactions

In organic synthesis, there are many examples of dehydration reaction, for example dehydration of alcohols or sugars.

Dehydration reactions		
Reaction	**Equation**	
Conversion of alcohols to ethers	2 R-OH → R-O-R + H_2O	
Conversion of alcohols to alkenes	R-CH_{2-} CHOH-R → R-CH=CH-R + H_2O	for example the conversion of glycerol to acrolein: 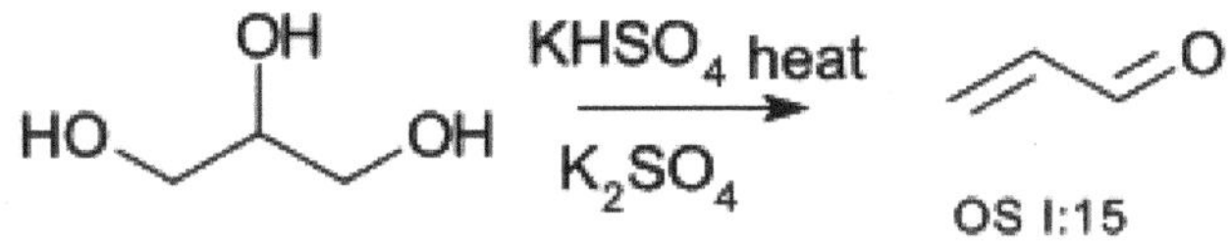 or the dehydration of 2-methyl-1-cyclohexanol to (mainly) 1-methylcyclohexene 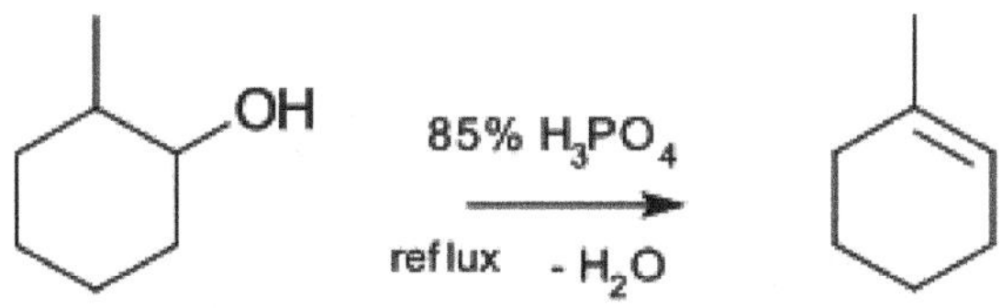
Conversion of carboxylic acids to acid anhydrides	2 RCOOH → $(RCO)_2O$ + H_2O	
Conversion of amides to nitriles	$RCONH_2$ → R-CN + H_2O	
Dienol benzene rearrangement		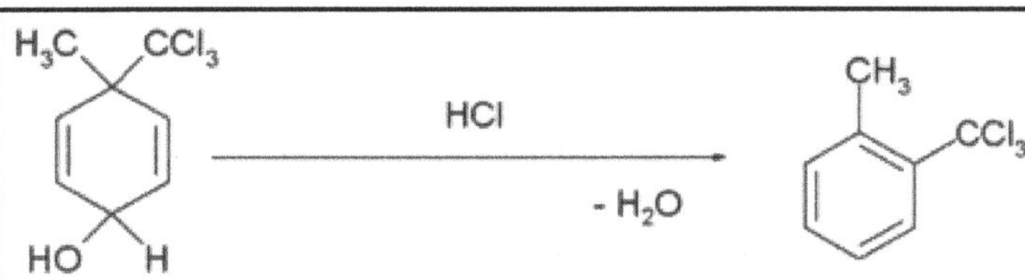

Some dehydration reactions can be mechanistically complex, for instance the reaction of a sugar (sucrose) with concentrated sulfuric acid: to form carbon as a graphitic foam involves formation of carbon-carbon bonds. The reaction is driven by the strongly exothermic reaction as sulfuric acid reacts with water, which produces dangerous sulfuric-acid containing steam, therefore the experiment should only be performed in a fume-hood or well ventilated area.

HALOGENATION

Halogenation is a chemical reaction that involves the addition of one or more halogens to a compound or material. Dehalogenation is the reverse of halogenation and results in the removal of a halogen from a molecule. The pathway and stoichiometry of halogenation depends on the structural features and functional groups of the organic substrate, as well as on the specific halogen. Inorganic compounds such as metals also undergo halogenation.

Organic Chemistry

Halogenation by Reaction Type

Several pathways exist for the halogenation of organic compounds, including free radical halogenation, ketone halogenation, electrophilic halogenation, and halogen addition reaction. The structure of the substrate is one factor that determines the pathway.

Free Radical Halogenation

Saturated hydrocarbons typically do not add halogens but undergo free radical halogenation, involving substitution of hydrogen atoms by halogen. The regiochemistry of the halogenation of alkanes is usually determined by the relative weakness of the available C–H bonds. The preference for reaction at tertiary and secondary positions results from greater stability of the corresponding free radicals and the transition state leading to them. Free radical halogenation is used for the industrial production of chlorinated methanes:

$CH_4 + Cl_2 \rightarrow CH_3Cl + HCl$

Rearrangement often accompany such free radical reactions.

Addition of Halogens to Alkenes and Alkynes

Unsaturated compounds, especially alkenes and alkynes, add halogens:

$RCH{=}CHR' + X_2 \rightarrow RCHX{-}CHXR'$

The addition of halogens to alkenes proceeds via intermediate halonium ions. In special cases, such intermediates have been isolated.

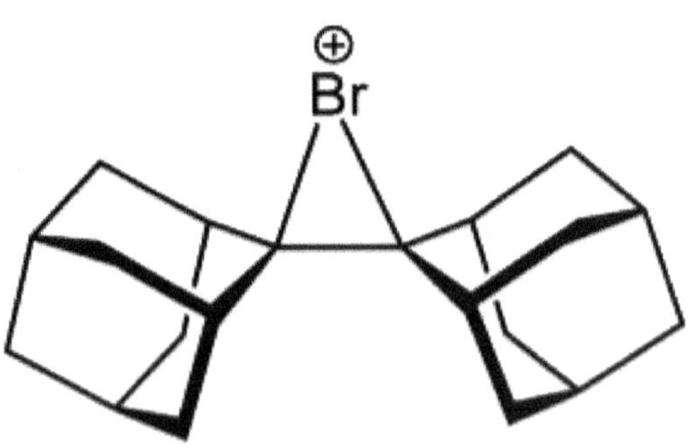

Structure of a bromonium ion.

Halogenation of Aromatic Compounds

Aromatic compounds are subject to electrophilic halogenation:

$RC_6H_5 + X_2 \rightarrow HX + RC_6H_4X$

The facility of halogenation is influenced by the halogen. Fluorine and chlorine are more electrophilic and are more aggressive halogenating agents. Bromine is a weaker halogenating agent than both fluorine and chlorine, while iodine is least reactive of them all. The facility of hydrogenolysis follows the reverse trend: iodine is most easily removed from organic compounds and organofluorine compounds are highly stable.

Other Halogenation Methods

In the Hunsdiecker reaction, from carboxylic acids are converted to the chain-shortened halide. The carboxylic acid is first converted to its silver salt, which is then oxidized with halogen:

$RCO_2Ag + Br_2 \rightarrow RBr + CO_2 + AgBr$

The Sandmeyer reaction is used to give aryl halides from diazonium salts, which are obtained from anilines.

In the Hell–Volhard–Zelinsky halogenation, carboxylic acids are alpha-halogenated.

In oxychlorination, the combination of hydrogen chloride and oxygen serves as the equivalent of chlorine, as illustrated by this route to dichloroethane:

$2\ HCl + CH_2{=}CH_2 + \frac{1}{2}\ O_2 \rightarrow ClCH_2CH_2Cl + H_2O$

Halogenation by Halogen Type

Fluorination

Organic compounds, saturated and unsaturated alike, react readily, usually explosively, with fluorine. Fluorination with elemental fluorine (F_2) requires highly specialised conditions and apparatus. Many commercially important organic compounds are fluorinated electrochemically using hydrogen fluoride as the source of fluorine. The method is called electrochemical fluorination. Aside from F_2 and its electrochemically generated equivalent, a variety of fluorinating reagents are known such as xenon difluoride and cobalt(III) fluoride.

Chlorination

Chlorination is generally highly exothermic. Both saturated and unsaturated compounds react directly with chlorine, the former usually requiring UV light to initiate homolysis of chlorine. Chlorination is conducted on a large scale industrially; major processes include routes to 1,2-dichloroethane (a precursor to PVC), as well as various chlorinated ethanes, as solvents.

Bromination

Bromination is more selective than chlorination because the reaction is less exothermic. Most commonly bromination is conducted by the addition of Br_2 to alkenes. An example of bromination is the organic synthesis of the anesthetic halothane from trichloroethylene:

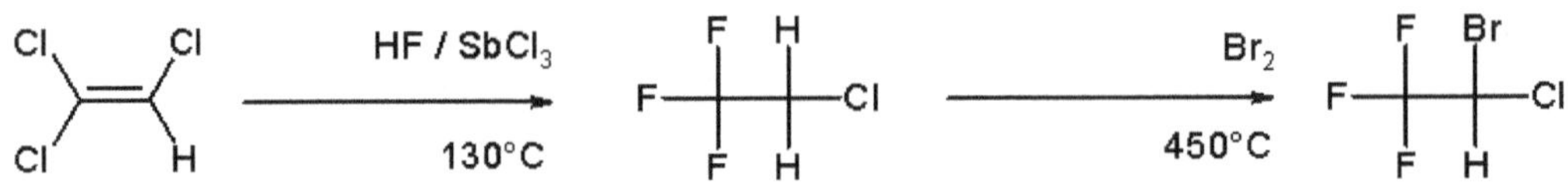

Organobromine compounds are the most common organohalides in nature. Their formation is catalyzed by the enzyme bromoperoxidase which utilizes bromide in combination with oxygen as an oxidant. The oceans are estimated to release 1–2 million tons of bromoform and 56,000 tons of bromomethane annually.

Iodination

Iodine is the least reactive halogen and is reluctant to react with most organic compounds. The addition of iodine to alkenes is the basis of the analytical method called the iodine number, a measure of the degree of unsaturation for fats. The iodoform reaction involves degradation of methyl ketones.

Inorganic Chemistry

All elements aside from argon, neon, and helium form fluorides by direct reaction with fluorine. Chlorine is slightly more selective, but still reacts with most metals and heavier nonmetals. Following the usual trend, bromine is less reactive and iodine least of all. Of the many reactions possible, illustrative is the formation of gold(III) chloride by the chlorination of gold. The chlorination of metals is usually not very important industrially since the chlorides are more easily made from the oxides and the hydrogen halide. Where chlorination of inorganic compounds is practiced on a relatively large scale is for the production of phosphorus trichloride and sulfur monochloride.

Chemical Dehalogenation

Chemical dehalogenation is a treatment to remove halogens from harmful chemicals or contaminated areas by making them less toxic. There are two types of dehalogenation: glycolate dehalogenation and base-catalyzed decomposition.

AROMATIC SULFONATION

Aromatic sulfonation is an organic reaction in which a hydrogen atom on an arene is replaced by a sulfonic acid functional group in an electrophilic aromatic substitution. Aryl sulfonic acids are used as detergents, dye, and drugs.

Stoichiometry and Mechanism

Typical conditions involve heating the aromatic compound with sulfuric acid:

$$C_6H_6 + H_2SO_4 \rightarrow C_6H_5SO_3H + H_2O$$

Sulfur trioxide or its protonated derivative is the actual electrophile in this electrophilic aromatic substitution.

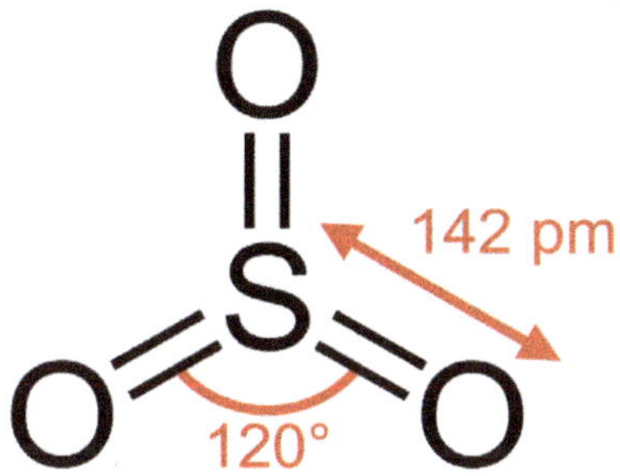

Sulfur trioxide is the active ingredient in many sulfonation reactions.

To drive the equilibrium, dehydrating agents such as thionyl chloride can be added.

$C_6H_6 + H_2SO_4 + SOCl_2 \rightarrow C_6H_5SO_3H + SO_2 + 2\ HCl$

Chlorosulfuric acid is also an effective agent:

$C_6H_6 + HSO_3Cl \rightarrow C_6H_5SO_3H + HCl$

In contrast to aromatic nitration and most other electrophilic aromatic substitutions this reaction is reversible. Sulfonation takes place in concentrated acidic conditions and desulfonation is the mode of action in a dilute hot aqueous acid. The reversibility is very useful in protecting the aromatic system.

Specialized Sulfonation Methods

Many method have been developed for introducing sulfonate groups aside from direction sulfonation.

Piria Reaction

A classic named reaction is the Piria reaction (R. Piria, 1851) in which nitrobenzene is reacted with a metal bisulfite forming an aminosulfonic acid as a result of combined nitro group reduction and sulfonation.

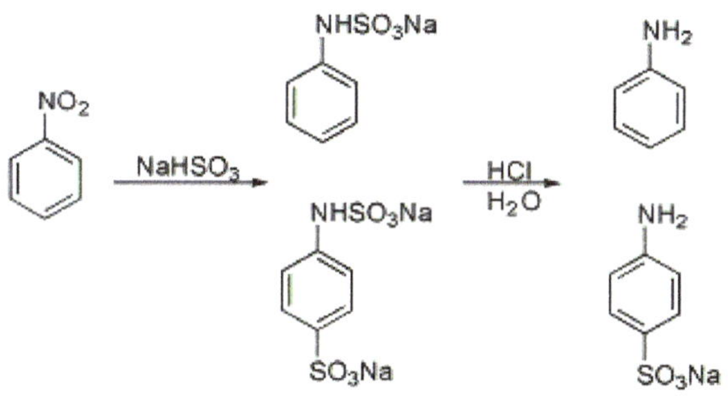

Tyre Sulfonation Process

In the Tyrer sulfonation process (1917), at some time of technological importance, benzene vapor is led through a vessel containing 90% sulfuric acid the temperature of which is increased from 100 to 180°C. Water and benzene are continuously removed in a condenser and the benzene layer fed back to the vessel. In this way an 80% yield is obtained.

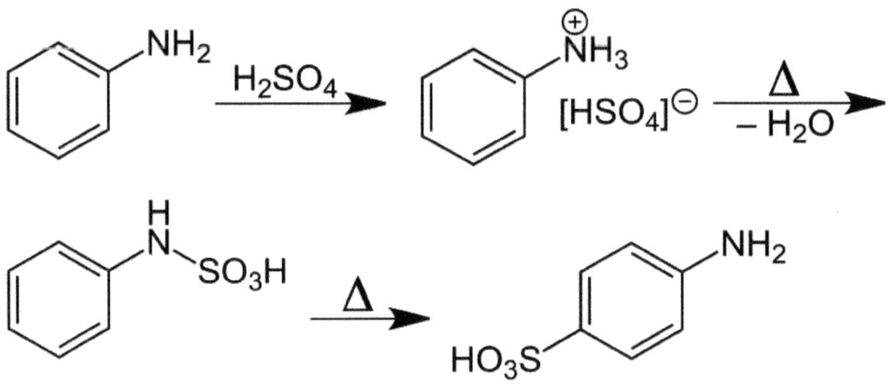

Synthesis of sulfanilic acid from aniline and sulfuric acid.

Applications

Aromatic sulfonic acids are intermediates in the preparation of dyes and many pharmaceuticals. Sulfonation of anilines lead to a large group of sulfa drugs.

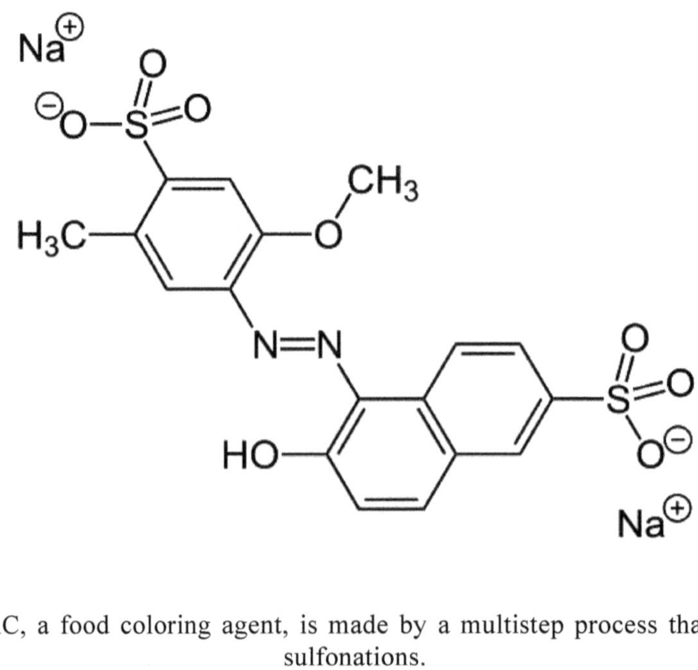

Allura Red AC, a food coloring agent, is made by a multistep process that includes two sulfonations.

Sulfonation of polystyrene is used to make sodium polystyrene sulfonate, a common ion exchange resin for water softening.

Reactions of Aryl Sulfonic acids

As a functional group, an aryl sulfonic acid exhibits two distinctive reactions:

- Desulfonation when heated in water near 200 °C. The reaction, a hydrolysis, is the reverse of the sulfonation. The temperature of desulfonation correlates with the ease of the sulfonation:

$RC_6H_4SO_3H + H_2O \rightarrow RC_6H_5 + H_2SO_4H$

This reactivity is exploited in the regiospecific conversion of 2-chlorotoluene by chlorination of p-toluenesulfonic acid, followed by hydrolysis of the intermediate.

- When treated with strong base, benzenesulfonic acid derivatives converts to phenols (via the phenoxides).

$C_6H_5SO_3H + 2\ NaOH \rightarrow C_6H_5ONa + NaHSO_4H$

Alkylation

Alkylation is the transfer of an alkyl group from one molecule to another. The alkyl group may be transferred as an alkyl carbocation, a free radical, a carbanion or a carbene (or their equivalents). An alkyl group is a piece of a molecule with the general formula C_nH_{2n+1}, where n is the integer depicting the number of carbons linked together. For example, a methyl group (n = 1, CH_3) is a fragment of a methane molecule (CH_4). Alkylating agents utilize selective alkylation by adding the desired aliphatic carbon chain to the previously chosen starting molecule. This is one of many known chemical syntheses. Alkyl groups can also be removed in a process known as dealkylation.

In oil refining contexts, alkylation refers to a particular alkylation of isobutane with olefins. For upgrading of petroleum, alkylation produces synthetic C_7–C_8 alkylate, which is a premium blending stock for gasoline.

In medicine, alkylation of DNA is used in chemotherapy to damage the DNA of cancer cells. Alkylation is accomplished with the class of drugs called alkylating antineoplastic agents.

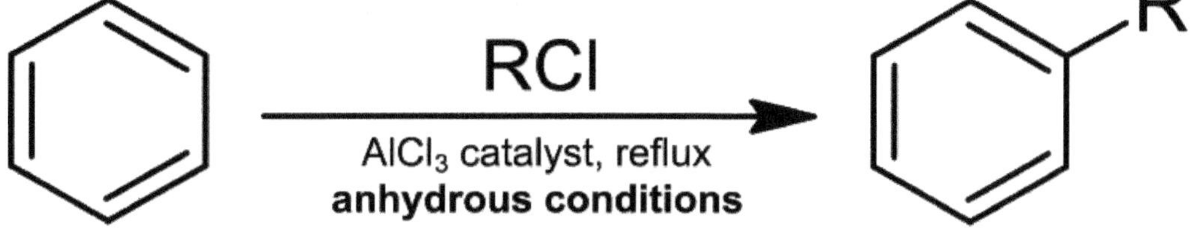

Benzene Friedel-Crafts alkylation.

Alkylating Agents

Alkylating agents are classified according to their nucleophilic or electrophilic character.

Nucleophilic Alkylating Agents

Nucleophilic alkylating agents deliver the equivalent of an alkyl anion (carbanion). Examples include the use of organometallic compounds such as Grignard (organomagnesium), organolithium, organocopper, and organosodium reagents. These compounds typically can add to an electron-deficient carbon atom such as at a carbonyl group. Nucleophilic alkylating agents can also displace halide substituents on a carbon atom. In the presence of catalysts, they also alkylate alkyl and aryl halides, as exemplified by Suzuki couplings.

Electrophilic Alkylating Agents

Electrophilic alkylating agents deliver the equivalent of an alkyl cation. Examples include the use of alkyl halides with a Lewis acid catalyst to alkylate aromatic substrates in Friedel-Crafts reactions. Alkyl halides can also react directly with amines to form C-N bonds; the same holds true for other nucleophiles such as alcohols, carboxylic acids, thiols, etc. Trimethyloxonium tetrafluoroborate and triethyloxonium tetrafluoroborate are particularly strong electrophiles due to their overt positive charge and an inert leaving group (dimethyl or diethyl ether).

Electrophilic, soluble alkylating agents are often very toxic, due to their ability to alkylate DNA. They should be handled with proper PPE. This mechanism of toxicity is also responsible for the ability of some alkylating agents to perform as anti-cancer drugs in the form of alkylating antineoplastic agents, and also as chemical weapons such as mustard gas. Alkylated DNA either does not coil or uncoil properly, or cannot be processed by information-decoding enzymes. This results in cytotoxicity with the effects of inhibition the growth of the cell, initiation of programmed cell death or apoptosis. However, mutations are also triggered, including carcinogenic mutations, explaining the higher incidence of cancer after exposure.

Alcohols and phenols can be alkylated to give alkyl ethers:

$$\mathrm{R{-}OH + R'{-}X \longrightarrow R{-}O{-}R' + H{-}X}$$

The produced acid HX is neutralized with a base, or, alternatively, the alcohol is deprotonated first to give an alkoxide or phenoxide. For example, dimethyl sulfate alkylates the sodium salt of phenol to give anisole, the methyl ether of phenol. The dimethyl sulfate is dealkylated to sodium methylsulfate.

$$\mathrm{Ph{-}O^- + Me_2{-}SO_4 \longrightarrow Ph{-}O{-}Me + Me{-}SO_4^-} \quad \text{(with } Na^+ \text{ as a spectator ion)}$$

On the contrary, the alkylation of amines introduces the problem that the alkylation of an amine makes it more nucleophilic. Thus, when an electrophilic alkylating agent is introduced to a primary amine, it will preferentially alkylate all the way to a quaternary ammonium cation.

$$\mathrm{R{-}NH_2 + R{-}NH{-}R' \longrightarrow R{-}N(R')_2 \longrightarrow R{-}N(R')_3^+} \quad \text{(alkylating agent omitted for clarity)}$$

If the quaternary ammonium is not the desired product, more circuitous routes such as reductive amination are necessary.

Carbene Alkylating Agents

Carbenes are extremely reactive and are known to attack even unactivated C-H bonds. Carbenes can be generated by elimination of a diazo group. Unlike electrophilic or nucleophilic alkylating agents, carbenes are neutral, and they insert into bonds rather than discard leaving groups. A metal can form a carbene equivalent called a transition metal carbene complex.

Catalysts

Silicotungstic acid is used to manufacture ethyl acetate by the alkylation of acetic acid by ethylene:

$C_2H_4 + CH_3CO_2H \rightarrow CH_3CO_2C_2H_5$

It has also been commercialized for the oxidation of ethylene to acetic acid:

$C_2H_4 + O_2 \rightarrow CH_3CO_2H$

In Biology

Methylation is the most common type of alkylation, being associated with the transfer of a methyl group. Methylation is distinct from alkylation in that it is specifically the transfer of one carbon, whereas alkylation can refer to the transfer of long chain carbon groups. Methylation in nature is typically effected by vitamin B12-derived enzymes, where the methyl group is carried by cobalt. In methanogenesis, coenzyme M is methylated by tetrahydromethanopterin.

Electrophilic compounds may alkylate different nucleophiles in the body. The toxicity, carcinogenity, and paradoxically, cancer cell-killing abilities of different DNA alkylating agents are an example.

Demethylation is the reverse of methylation.

Oil Refining

In a standard oil refinery process, isobutane is alkylated with low-molecular-weight alkenes (primarily a mixture of propene and butene) in the presence of a Bronsted acid catalyst, either sulfuric acid or hydrofluoric acid. In an oil refinery it is referred to as a sulfuric acid alkylation unit (SAAU) or a hydrofluoric alkylation unit, (HFAU). Refinery workers may simply refer to it as the alky or alky unit. The catalyst protonates the alkenes (propene, butene) to produce reactive carbocations, which alkylate isobutane. The reaction is carried out at mild temperatures (0 and 30 °C) in a two-phase reaction. Because the reaction is exothermic, cooling is needed: SAAU plants require lower temperatures so the cooling medium needs to be chilled, for HFAU normal refinery cooling water will suffice. It is important to keep a high ratio of isobutane to alkene at the point of reaction to prevent side reactions which produces a lower octane product, so the plants have a high recycle of isobutane back to feed. The phases separate spontaneously, so the acid phase is vigorously mixed with the hydrocarbon phase to create sufficient contact surface.

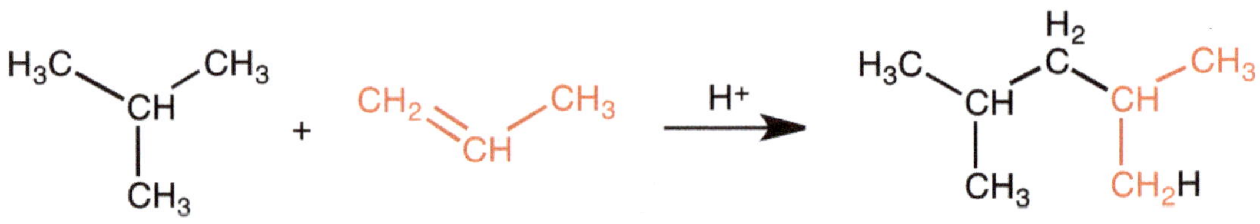

Alkylation of alkenes (shown in red is propene) by isobutane is a major process in refineries to produce higher octane "alkylate" for gasoline blending, in this example yielding isoheptane. It is catalysed by strong acids such as hydrofluoric acid (HF) and sulfuric acid (H_2SO_4).

The product is called alkylate and is composed of a mixture of high-octane, branched-chain paraffinic hydrocarbons (mostly isoheptane and isooctane). Alkylate is a premium gasoline blending stock because it has exceptional antiknock properties and is clean burning. Alkylate is also a key component of avgas. The octane number of the alkylate depends mainly upon the kind of alkenes used and upon operating conditions. For example, isooctane results from combining butylene with isobutane and has an octane rating of 100 by definition. There are other products in the alkylate, so the octane rating will vary accordingly.

Since crude oil generally contains only 10 to 40 percent of hydrocarbon constituents in the gasoline range, refineries use a fluid catalytic cracking process to convert high molecular weight hydrocarbons into smaller and more volatile compounds, which are then converted into liquid gasoline-size hydrocarbons. Alkylation processes transform low molecular-weight alkenes and iso-paraffin molecules into larger iso-paraffins with a high octane number.

Combining cracking, polymerization, and alkylation can result in a gasoline yield representing 70 percent of the starting crude oil. More advanced processes, such as cyclicization of paraffins and dehydrogenation of naphthenes forming aromatic hydrocarbons in a catalytic reformer, have also been developed to increase the octane rating of gasoline. Modern refinery operation can be shifted to produce almost any fuel type with specified performance criteria from a single crude feedstock.

Refineries examine whether it makes sense economically to install alkylation units. Alkylation units are complex, with substantial economy of scale. In addition to a suitable quantity of feedstock, the price spread between the value of alkylate product and alternate feedstock disposition value must be large enough to justify the installation. Alternative outlets for refinery alklylation feedstocks include sales as LPG, blending of C4 streams directly into gasoline to lower the flash point of the product and feedstocks for chemical plants. Local market conditions vary widely between plants. Variation in the RVP (Reid vapor pressure) specification for gasoline between countries and between seasons dramatically impacts the amount of butane streams that can be blended directly into gasoline. The transportation of specific types of LPG streams can be expensive so local disparities in economic conditions are often not fully mitigated by cross market movements of alkylation feedstocks.

The availability of a suitable catalyst is also an important factor in deciding whether to build an alkylation plant. If sulfuric acid is used, significant volumes are needed. Access to a suitable plant is required for the supply of fresh acid and the disposition of spent acid. If a sulfuric acid plant must be constructed specifically to support an alkylation unit, such construction will have a significant impact on both the initial requirements for capital and ongoing costs of operation. Alternatively it is possible to install a WSA Process unit to regenerate the spent acid. No drying of the gas takes place. This means that there will be no loss of acid, no acidic waste material and no heat is lost in process gas reheating. The selective condensation in the WSA condenser ensures that the regenerated fresh acid will be 98% w/w even with the humid process gas. It is possible to combine spent acid regeneration with disposal of hydrogen sulfide by using the hydrogen sulfide as internal fuel in the refinery or elsewhere.

The second main catalyst option is hydrofluoric acid. In typical alkylation plants, rates of consumption for acid are much lower than for sulfuric acid. These plants also produce alkylate with better octane rating than do sulfuric plants. However, due to its hazardous nature, HF acid is produced at very few locations and transportation must be managed rigorously.

ESTER

In chemistry, esters are chemical compounds derived from an acid (organic or inorganic) in which at least one –OH (hydroxyl) group is replaced by an –O–alkyl (alkoxy) group. Usually, esters are derived from a carboxylic acid and an alcohol. Glycerides, which are fatty acid esters of glycerol, are important esters in biology, being one of the main classes of lipids, and making up the bulk of animal fats and vegetable oils. Esters with low molecular weight are commonly used as fragrances and found in essential oils and pheromones. Phosphoesters form the backbone of DNA molecules. Nitrate esters, such as nitroglycerin, are known for their explosive properties, while polyesters are important plastics, with monomers linked by ester moieties.

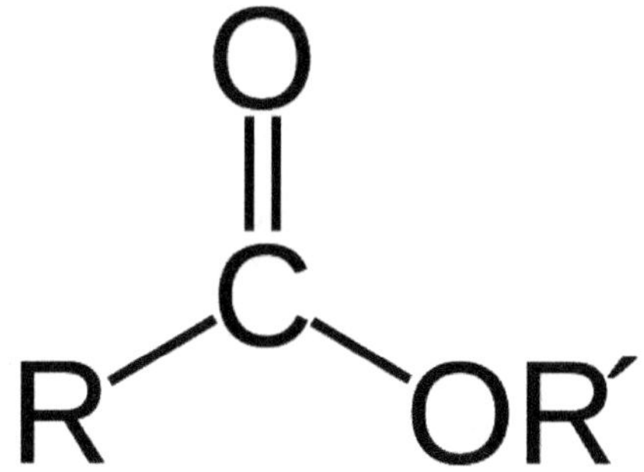

A carboxylate ester. R and R′ denote any alkyl or aryl group. R can also be a hydrogen atom.

Nomenclature

Etymology

The word 'ester' was coined in 1848 by German chemist Leopold Gmelin, probably as a contraction of the German Essigäther, "acetic ether".

IUPAC Nomenclature

Main article: IUPAC nomenclature of organic chemistry § Esters

Ester names are derived from the parent alcohol and the parent acid, where the latter may be organic or inorganic. Esters derived from the simplest carboxylic acids are commonly named according to the more traditional, so-called "trivial names" e.g. as formate, acetate, propionate, and butyrate, as opposed to the IUPAC nomenclature methanoate, ethanoate, propanoate and butanoate. Esters derived from more complex carboxylic acids are, on the other hand, more frequently named using the systematic IUPAC name, based on the name for the acid followed by the suffix -oate. For example, the ester hexyl octanoate, also known under the trivial name hexyl caprylate, has the formula $CH_3(CH_2)_6CO_2(CH_2)_5CH_3$.

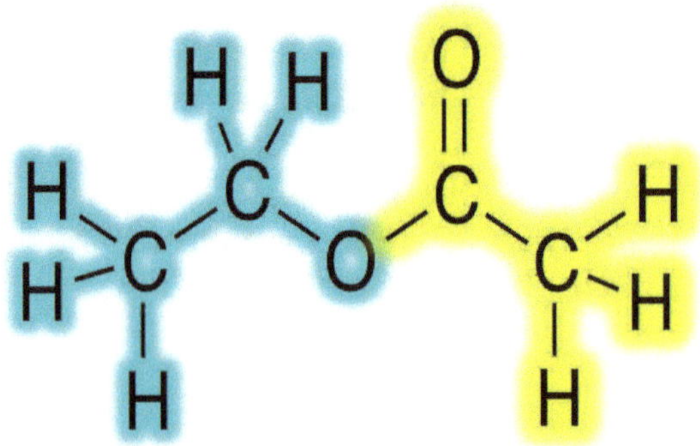

Ethyl acetate derived from an alcohol (blue) and an acyl group (yellow) derived from a carboxylic acid.

The chemical formulas of organic esters usually take the form RCO_2R', where R and R′ are the hydrocarbon parts of the carboxylic acid and the alcohol, respectively. For example, butyl acetate (systematically butyl ethanoate), derived from butanol and acetic acid (systematically ethanoic acid) would be written $CH_3CO_2C_4H_9$. Alternative presentations are common including BuOAc and $CH_3COOC_4H_9$.

Cyclic esters are called lactones, regardless of whether they are derived from an organic or an inorganic acid. One example of a (organic) lactone is γ-valerolactone.

Orthoesters

An uncommon class of organic esters are the orthoesters, which have the formula $RC(OR')_3$. Triethylorthoformate ($HC(OC_2H_5)_3$) is derived, in terms of its name (but not its synthesis) from orthoformic acid ($HC(OH)_3$) and ethanol.

Inorganic Esters

Esters can also be derived from an inorganic acid and an alcohol. Thus, the nomenclature extends to inorganic oxo acids and their corresponding esters: phosphoric acid and phosphate esters/organophosphates, sulfuric acid and sulfate esters/organosulfates, nitric acid and nitrate, and boric acid and borates. For example, triphenyl phosphate is the ester derived from phosphoric acid and phenol. Organic carbonates are derived from carbonic acid; for example, ethylene carbonate is derived from carbonic acid and ethylene glycol.

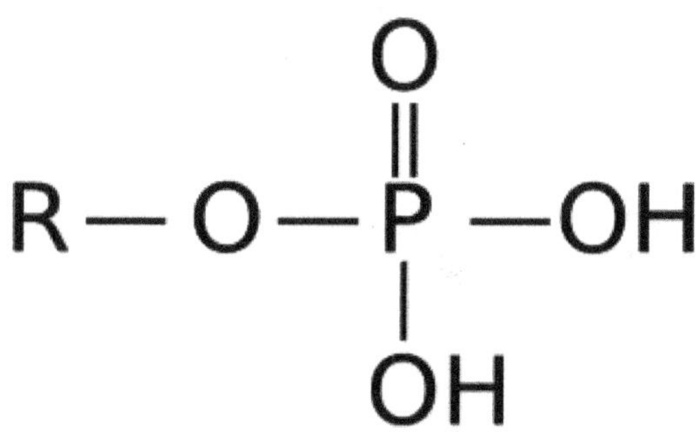

A phosphoric acid ester.

So far an alcohol and inorganic acid are linked via oxygen atoms. The definition of inorganic acid ester that feature inorganic chemical elements links between alcohols and the inorganic acid – the phosphorus atom linking to three alkoxy functional groups in organophosphate – can be extended to the same elements in various combinations of covalent bonds between carbons and the central inorganic atom and carbon–oxygen bonds to central inorganic atoms. For example, phosphorus features three carbon–oxygen–phosphorus bondings and one phosphorus–oxygen double bond in organophosphates,

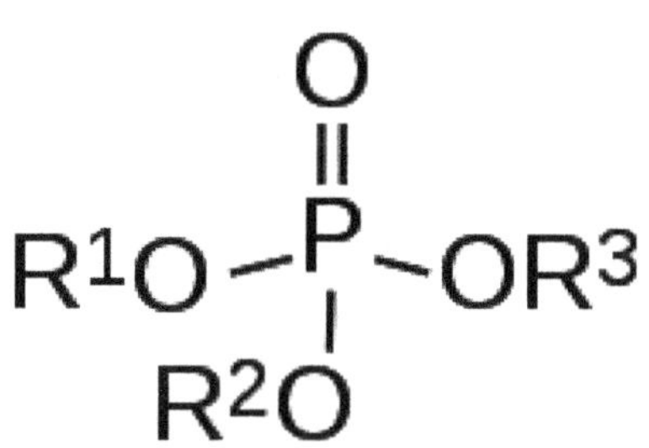

Structure of a generic organophosphate.

Three carbon–oxygen–phosphorus bondings and no phosphorus–oxygen double bonds in phosphite esters or organophosphites,

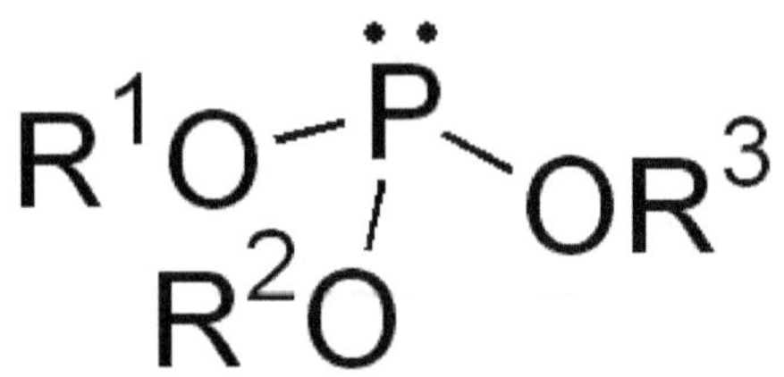

Structure of a generic phosphite ester showing the lone pairs on the P.

Two carbon–oxygen–phosphorus bondings, no phosphorus–oxygen double bonds but one phosphorus–carbon bond in phosphonites,

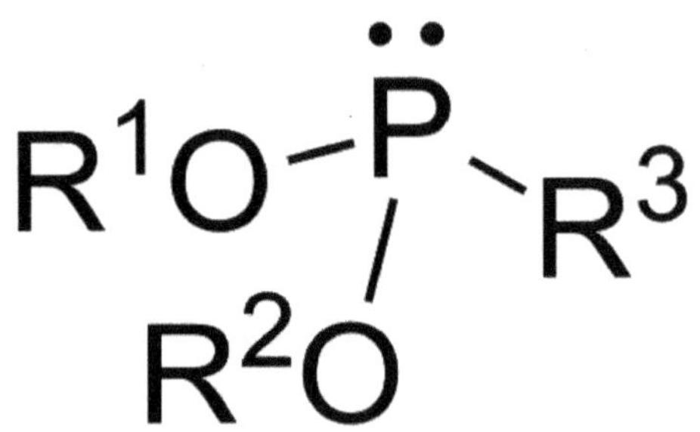

Structure of a generic phosphonite – ester of phosphonous acid.

One carbon–oxygen–phosphorus bondings, no phosphorus–oxygen double bonds but two phosphorus–carbon bonds in phosphinites.

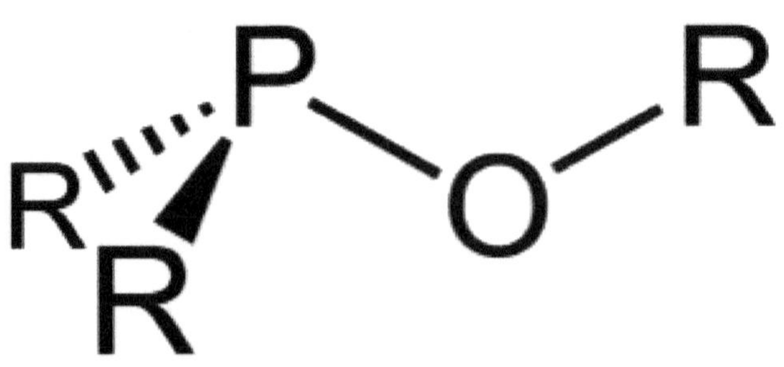

Structure of a generic phosphinite.

In corollary, boron features borinic esters (n = 2), boronic esters (n = 1), and borates (n = 0).

As oxygen is a group 16 chemical element, sulfur atoms can replace some oxygen atoms in carbon–oxygen–central inorganic atom covalent bonds of an ester. As a result, thiosulfinates' and thiosulfonates, with a central inorganic sulfur atom, demonstrate clearly the assortment of sulfur esters, that also includes sulfates, sulfites, sulfonates, sulfinates, sulfenates esters.

Structure and Bonding

Esters contain a carbonyl center, which gives rise to 120 ° C–C–O and O–C–O angles. Unlike amides, esters are structurally flexible functional groups because rotation about the C–O–C bonds has a low barrier. Their flexibility and low polarity is manifested in their physical properties; they tend to be less rigid (lower melting point) and more volatile (lower boiling point) than the corresponding amides. The pK_a of the alpha-hydrogens on esters is around 25.

Many esters have the potential for conformational isomerism, but they tend to adopt an s-cis (or Z) conformation rather than the s-trans (or E) alternative, due to a combination of hyperconjugation and dipole minimization effects. The preference for the Z conformation is influenced by the nature of the substituents and solvent, if present. Lactones with small rings are restricted to the s-trans (i.e. E) conformation due to their cyclic structure.

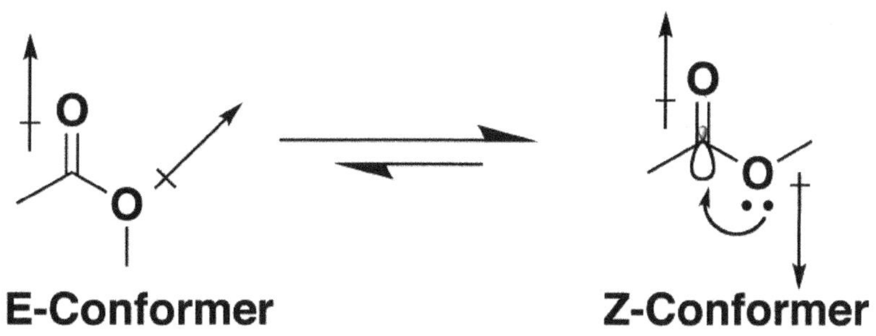

Physical Properties and Characterization

Esters are more polar than ethers but less polar than alcohols. They participate in hydrogen bonds as hydrogen-bond acceptors, but cannot act as hydrogen-bond donors, unlike their parent alcohols. This ability to participate in hydrogen bonding confers some water-solubility. Because of their lack of hydrogen-bond-donating ability, esters do not self-associate. Consequently, esters are more volatile than carboxylic acids of similar molecular weight.

Characterization and Analysis

Esters are generally identified by gas chromatography, taking advantage of their volatility. IR spectra for esters feature an intense sharp band in the range 1730–1750 cm^{-1} assigned to $\nu_{C=O}$. This peak changes depending on the functional groups attached to the carbonyl.

For example, a benzene ring or double bond in conjugation with the carbonyl will bring the wavenumber down about 30 cm^{-1}.

Applications and Occurrence

Esters are widespread in nature and are widely used in industry. In nature, fats are in general triesters derived from glycerol and fatty acids. Esters are responsible for the aroma of many fruits, including apples, durians, pears, bananas, pineapples, and strawberries. Several billion kilograms of polyesters are produced industrially annually, important products being polyethylene terephthalate, acrylate esters, and cellulose acetate.

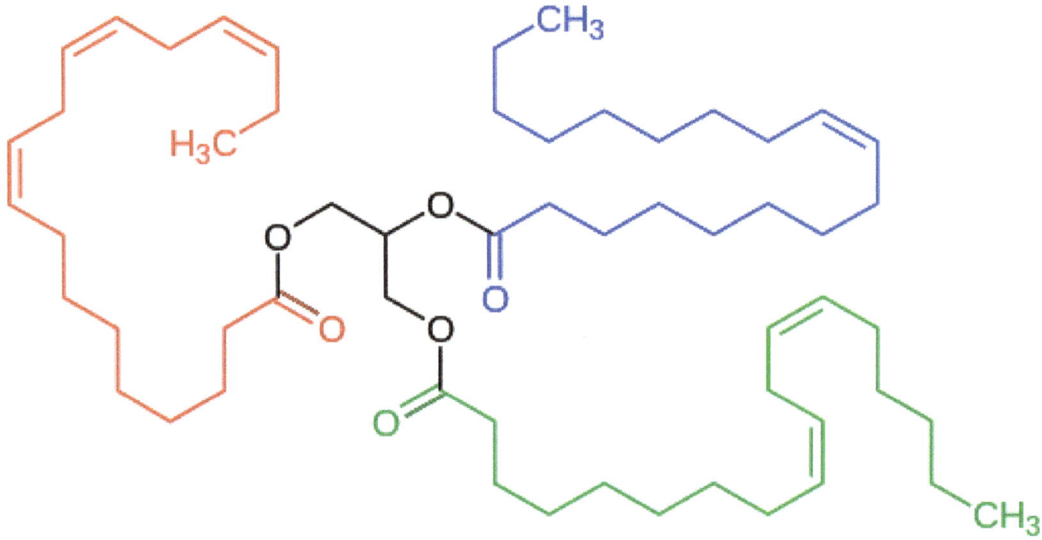

Representative triglyceride found in a linseed oil, a triester (triglyceride) derived of linoleic acid, alpha-linolenic acid, and oleic acid.

Preparation

Esterification is the general name for a chemical reaction in which two reactants (typically an alcohol and an acid) form an ester as the reaction product. Esters are common in organic chemistry and biological materials, and often have a characteristic pleasant, fruity odor. This leads to their extensive use in the fragrance and flavor industry. Ester bonds are also found in many polymers.

Esterification of Carboxylic Acids

The classic synthesis is the Fischer esterification, which involves treating a carboxylic acid with an alcohol in the presence of a dehydrating agent:

$$RCO_2H + R'OH \rightleftharpoons RCO_2R' + H_2O$$

The equilibrium constant for such reactions is about 5 for typical esters, e.g., ethyl acetate. The reaction is slow in the absence of a catalyst. Sulfuric acid is a typical catalyst for this reaction. Many other acids are also used such as polymeric sulfonic acids. Since esterification is highly reversible, the yield of the ester can be improved using Le Chatelier's principle:

- Using the alcohol in large excess (i.e., as a solvent).
- Using a dehydrating agent: sulfuric acid not only catalyzes the reaction but sequesters water (a reaction product). Other drying agents such as molecular sieves are also effective.
- Removal of water by physical means such as distillation as a low-boiling azeotropes with toluene, in conjunction with a Dean-Stark apparatus.

Reagents are known that drive the dehydration of mixtures of alcohols and carboxylic acids. One example is the Steglich esterification, which is a method of forming esters under mild conditions. The method is popular in peptide synthesis, where the substrates are sensitive to harsh conditions like high heat. DCC (dicyclohexylcarbodiimide) is used to activate the carboxylic acid to further reaction. DMAP (4-dimethylaminopyridine) is used as an acyl-transfer catalyst.

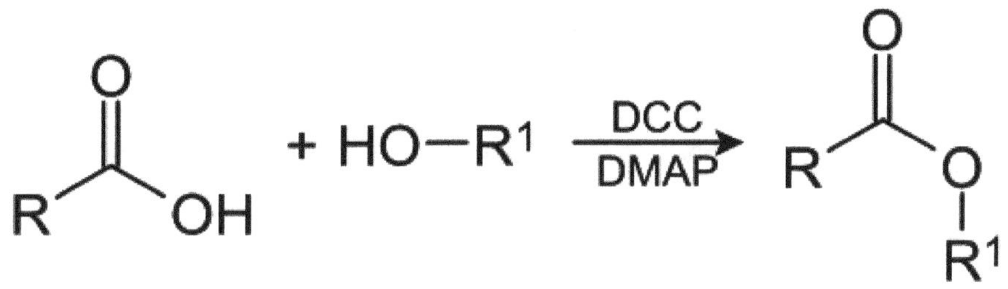

Another method for the dehydration of mixtures of alcohols and carboxylic acids is the Mitsunobu reaction:

$RCO_2H + R'OH + P(C_6H_5)_3 + R_2N_2 \rightarrow RCO_2R' + OP(C_6H_5)_3 + R_2N_2H_2$

Carboxylic acids can be esterified using diazomethane:

$RCO_2H + CH_2N_2 \rightarrow RCO_2CH_3 + N_2$

Using this diazomethane, mixtures of carboxylic acids can be converted to their methyl esters in near quantitative yields, e.g., for analysis by gas chromatography. The method is useful in specialized organic synthetic operations but is considered too hazardous and expensive for large scale applications.

Alcoholysis of Acyl Chlorides and Acid Anhydrides

Alcohols react with acyl chlorides and acid anhydrides to give esters:

$RCOCl + R'OH \rightarrow RCO_2R' + HCl$

$(RCO)_2O + R'OH \rightarrow RCO_2R' + RCO_2H$

The reactions are irreversible simplifying work-up. Since acyl chlorides and acid anhydrides also react with water, anhydrous conditions are preferred. The analogous acylations of amines to give amides are less sensitive because amines are stronger nucleophiles and react more rapidly than does water. This method is employed only for laboratory-scale procedures, as it is expensive.

Alkylation of Carboxylate Salts

Although not widely employed for esterifications, salts of carboxylate anions can be alkylating agent with alkyl halides to give esters. In the case that an alkyl chloride is used, an iodide salt can catalyze the reaction (Finkelstein reaction). The carboxylate salt is often generated in situ. In difficult cases, the silver carboxylate may be used, since the silver ion coordinates to the halide aiding its departure and improving the reaction rate. This reaction can suffer from anion availability problems and, therefore, can benefit from the addition of phase transfer catalysts or highly polar aprotic solvents such as DMF.

Transesterification

Transesterification, which involves changing one ester into another one, is widely practiced:

$$RCO_2R' + CH_3OH \rightarrow RCO_2CH_3 + R'OH$$

Like the hydrolysation, transesterification is catalysed by acids and bases. The reaction is widely used for degrading triglycerides, e.g. in the production of fatty acid esters and alcohols. Poly(ethylene terephthalate) is produced by the transesterification of dimethyl terephthalate and ethylene glycol:

$$(C_6H_4)(CO_2CH_3)_2 + 2\ C_2H_4(OH)_2 \rightarrow \tfrac{1}{n}\ \{(C_6H_4)(CO_2)_2(C_2H_4)\}_n + 2\ CH_3OH$$

Carbonylation

Alkenes undergo "hydroesterification" in the presence of metal carbonyl catalysts. Esters of propionic acid are produced commercially by this method:

$$C_2H_4 + ROH + CO \rightarrow C_2H_5CO_2R$$

The carbonylation of methanol yields methyl formate, which is the main commercial source of formic acid. The reaction is catalyzed by sodium methoxide:

$$CH_3OH + CO \rightarrow CH_3O_2CH$$

Addition of Carboxylic Acids to Alkenes

In the presence of palladium-based catalysts, ethylene, acetic acid, and oxygen react to give vinyl acetate:

$$C_2H_4 + CH_3CO_2H + \tfrac{1}{2}\ O_2 \rightarrow C_2H_3O_2CCH_3 + H_2O$$

Direct routes to this same ester are not possible because vinyl alcohol is unstable.

Other Methods

- Favorskii rearrangement of α-haloketones in presence of base
- Baeyer–Villiger oxidation of ketones with peroxides
- Pinner reaction of nitriles with an alcohol
- Nucleophilic abstraction of a metal–acyl complex
- Hydrolysis of orthoesters in aqueous acid
- Cellulolysis via esterification
- Ozonolysis of alkenes using a work up in the presence of hydrochloric acid and various alcohols.
- Anodic oxidation of methyl ketones leading to methyl esters.

Reactions

Esters react with nucleophiles at the carbonyl carbon. The carbonyl is weakly electrophilic but is attacked by strong nucleophiles (amines, alkoxides, hydride sources, organolithium compounds, etc.). The C–H bonds adjacent to the carbonyl are weakly acidic but undergo deprotonation with strong bases. This process is the one that usually initiates condensation reactions. The carbonyl oxygen is weakly basic (less so than in amides) but forms adducts.

Addition of Nucleophiles at Carbonyl

Esterification is a reversible reaction. Esters undergo hydrolysis under acid and basic conditions. Under acidic conditions, the reaction is the reverse reaction of the Fischer esterification. Under basic conditions, hydroxide acts as a nucleophile, while an alkoxide is the leaving group. This reaction, saponification, is the basis of soap making.

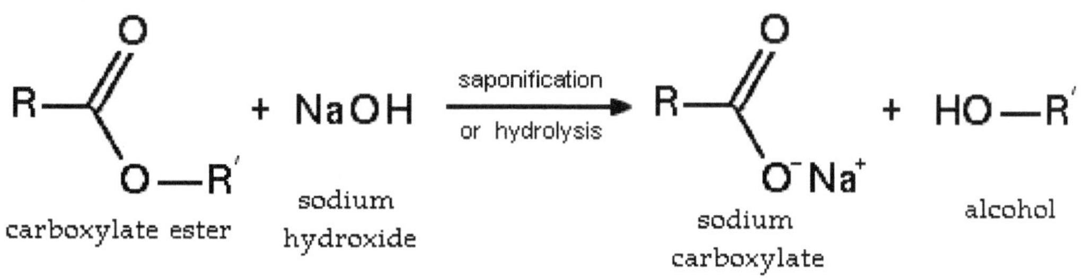

The alkoxide group may also be displaced by stronger nucleophiles such as ammonia or primary or secondary amines to give amides: (ammonolysis reaction).

$RCO_2R' + NH_2R'' \rightarrow RCONHR'' + R'OH$

This reaction is not usually reversible. Hydrazines and hydroxylamine can be used in place of amines. Esters can be converted to isocyanates through intermediate hydroxamic acids in the Lossen rearrangement.

Sources of carbon nucleophiles, e.g., Grignard reagents and organolithium compounds, add readily to the carbonyl.

Reduction

Compared to ketones and aldehydes, esters are relatively resistant to reduction. The introduction of catalytic hydrogenation in the early part of the 20th century was a breakthrough; esters of fatty acids are hydrogenated to fatty alcohols.

$RCO_2R' + 2\ H_2 \rightarrow RCH_2OH + R'OH$

A typical catalyst is copper chromite. Prior to the development of catalytic hydrogenation, esters were reduced on a large scale using the Bouveault–Blanc reduction. This method, which is largely obsolete, uses sodium in the presence of proton sources.

Especially for fine chemical syntheses, lithium aluminium hydride is used to reduce esters to two primary alcohols. The related reagent sodium borohydride is slow in this reaction. DIBAH reduces esters to aldehydes.

Direct reduction to give the corresponding ether is difficult as the intermediate hemiacetal tends to decompose to give an alcohol and an aldehyde (which is rapidly reduced to give a second alcohol). The reaction can be achieved using triethylsilane with a variety of Lewis acids.

Claisen Condensation and Related Reactions

As for aldehydes, the hydrogen atoms on the carbon adjacent ("α to") the carboxyl group in esters are sufficiently acidic to undergo deprotonation, which in turn leads to a variety of useful reactions. Deprotonation requires relatively strong bases, such as alkoxides. Deprotonation gives a nucleophilic enolate, which can further react, e.g., the Claisen condensation and its intramolecular equivalent, the Dieckmann condensation. This conversion is exploited in the malonic ester synthesis, wherein the diester of malonic acid reacts with an electrophile (e.g., alkyl halide), and is subsequently decarboxylated. Another variation is the Fráter–Seebach alkylation.

Other Reactions

- Phenyl esters react to hydroxyarylketones in the Fries rearrangement.
- Specific esters are functionalized with an α-hydroxyl group in the Chan rearrangement.
- Esters with β-hydrogen atoms can be converted to alkenes in ester pyrolysis.
- A direct conversion of esters to nitriles.

Protecting Groups

As a class, esters serve as protecting groups for carboxylic acids. Protecting a carboxylic acid is useful in peptide synthesis, to prevent self-reactions of the bifunctional amino acids. Methyl and ethyl esters are commonly available for many amino acids; the t-butyl ester tends to be more expensive. However, t-butyl esters are particularly useful because, under strongly acidic conditions, the t-butyl esters undergo elimination to give the carboxylic acid and isobutylene, simplifying work-up.

List of Ester Odorants

Many esters have distinctive fruit-like odors, and many occur naturally in the essential oils of plants. This has also led to their commonplace use in artificial flavorings and fragrances when those odors aim to be mimicked.

Ester name	Formula	Odor or occurrence
Allyl hexanoate	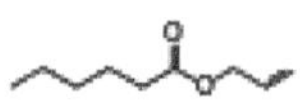	pineapple
Benzyl acetate	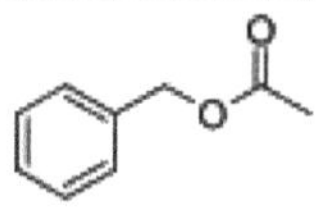	pear, strawberry, jasmine
Bornyl acetate	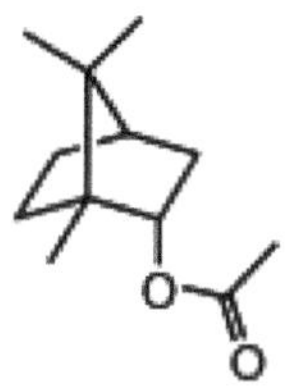	pine
Butyl acetate	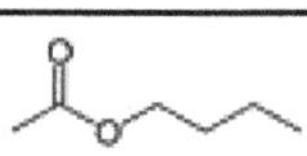	apple, honey
Butyl butyrate	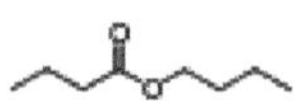	pineapple
Butyl propanoate		pear drops
Ethyl acetate	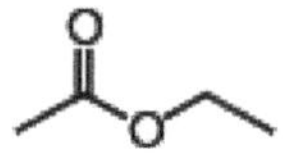	nail polish remover, model paint, model airplane glue
Ethyl benzoate	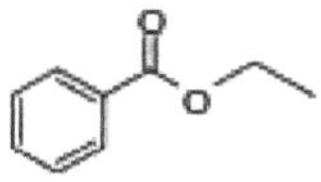	sweet, wintergreen, fruity, medicinal, cherry, grape
Ethyl butyrate	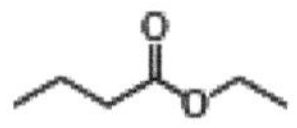	banana, pineapple, strawberry
Ethyl hexanoate	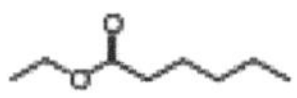	pineapple, waxy-green banana
Ethyl cinnamate	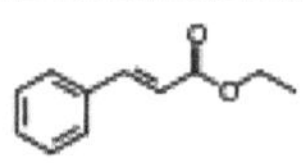	cinnamon
Ethyl formate		lemon, rum, strawberry

Ethyl heptanoate		apricot, cherry, grape, raspberry
Ethyl isovalerate	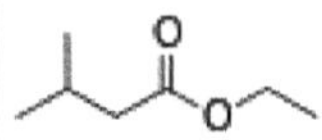	apple
Ethyl lactate	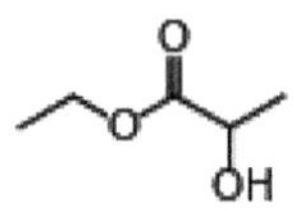	butter, cream
Ethyl nonanoate		grape
Ethyl pentanoate	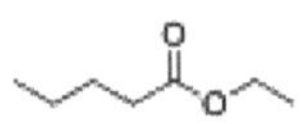	apple
Geranyl acetate	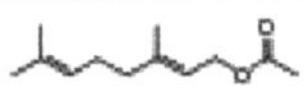	geranium
Geranyl butyrate		cherry
Geranyl pentanoate	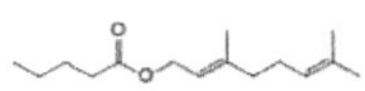	apple
Isobutyl acetate	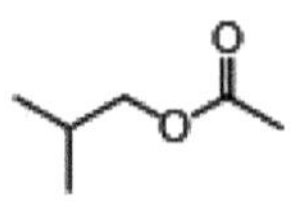	cherry, raspberry, strawberry
Isobutyl formate	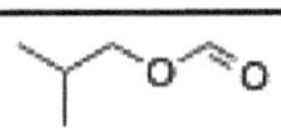	raspberry
Isoamyl acetate	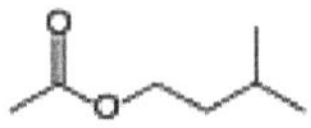	pear, banana (flavoring in Pear drops)
Isopropyl acetate	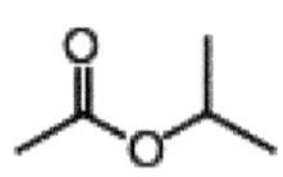	fruity
Linalyl acetate	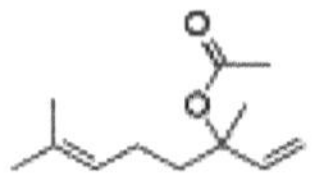	lavender, sage
Linalyl butyrate	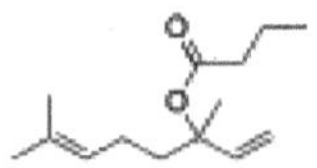	peach
Linalyl formate	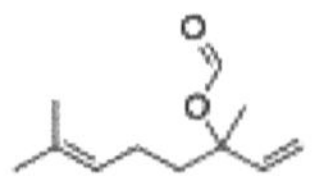	apple, peach
Methyl acetate	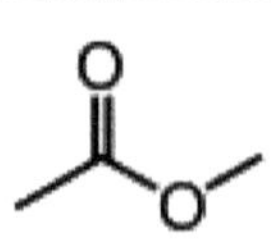	glue
Methyl anthranilate	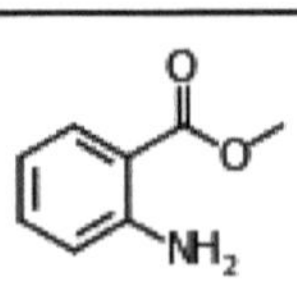	grape, jasmine

Methyl benzoate	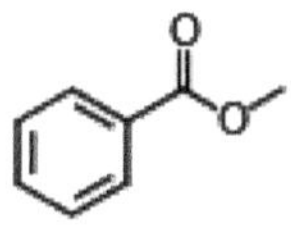	fruity, ylang ylang, feijoa
Methyl butyrate (methyl butanoate)	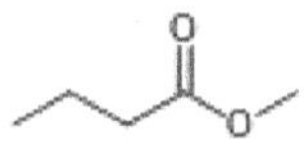	pineapple, apple, strawberry
Methyl cinnamate	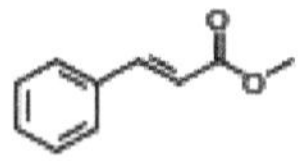	strawberry
Methyl pentanoate (methyl valerate)	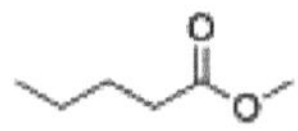	flowery
Methyl phenylacetate	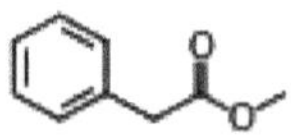	honey
Methyl salicylate (oil of wintergreen)	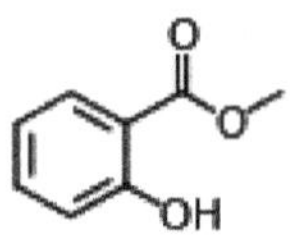	Modern root beer, wintergreen, Germolene and Ralgex ointments (UK)
Nonyl caprylate	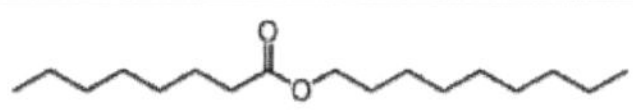	orange
Octyl acetate	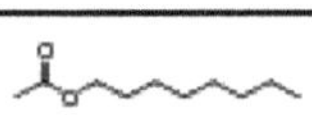	fruity-orange
Octyl butyrate	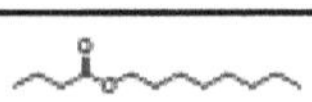	parsnip
Amyl acetate (pentyl acetate)	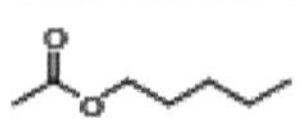	apple, banana
Pentyl butyrate (amyl butyrate)	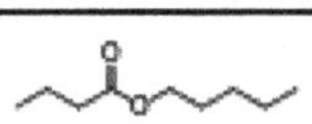	apricot, pear, pineapple
Pentyl hexanoate (amyl caproate)	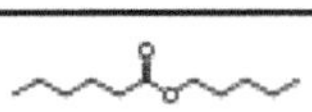	apple, pineapple
Pentyl pentanoate (amyl valerate)	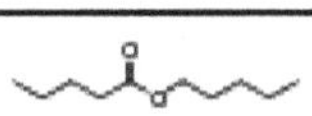	apple
Propyl acetate	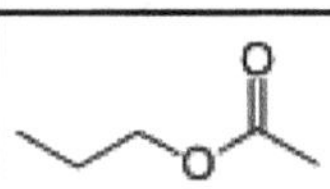	pear
Propyl hexanoate	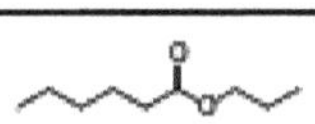	blackberry, pineapple, cheese, wine
Propyl isobutyrate	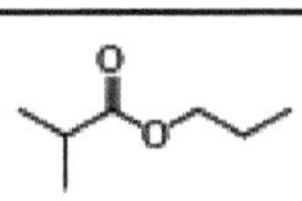	rum
Terpenyl butyrate	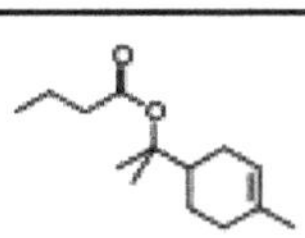	cherry

POLYMERIZATION

In polymer chemistry, polymerization is a process of reacting monomer molecules together in a chemical reaction to form polymer chains or three-dimensional networks. There are many forms of polymerization and different systems exist to categorize them.

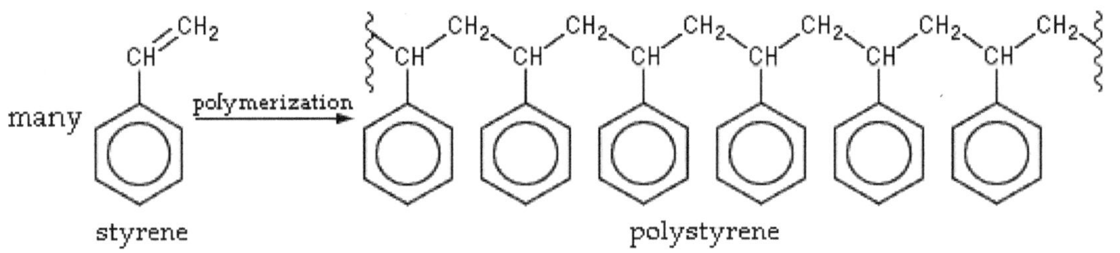

An example of **alkene polymerization**, in which each styrene monomer's double bond reforms as a single bond plus a bond to another styrene monomer. The product is polystyrene.

Introduction

In chemical compounds, polymerization occurs via a variety of reaction mechanisms that vary in complexity due to functional groups present in reacting compounds and their inherent steric effects. In more straightforward polymerization, alkenes, which are relatively stable due to σ bonding between carbon atoms, form polymers through relatively simple radical reactions; in contrast, more complex reactions such as those that involve substitution at the carbonyl group require more complex synthesis due to the way in which reacting molecules polymerize. Alkanes can also be polymerized, but only with the help of strong acids.

As alkenes can be formed in somewhat straightforward reaction mechanisms, they form useful compounds such as polyethylene and polyvinyl chloride (PVC) when undergoing radical reactions, which are produced in high tonnages each year due to their usefulness in manufacturing processes of commercial products, such as piping, insulation and packaging. In general, polymers such as PVC are referred to as "homopolymers," as they consist of repeated long chains or structures of the same monomer unit, whereas polymers that consist of more than one molecule are referred to as copolymers (or co-polymers).

Other monomer units, such as formaldehyde hydrates or simple aldehydes, are able to polymerize themselves at quite low temperatures (ca. −80 °C) to form trimers; molecules consisting of 3 monomer units, which can cyclize to form ring cyclic structures, or undergo further reactions to form tetramers, or 4 monomer-unit compounds. Further compounds either being referred to as oligomers in smaller molecules. Generally, because formaldehyde is an exceptionally reactive electrophile it allows nucleophillic addition of hemiacetal intermediates, which are in general short-lived and relatively unstable «mid-stage» compounds that react with other molecules present to form more stable polymeric compounds.

Polymerization that is not sufficiently moderated and proceeds at a fast rate can be very hazardous. This phenomenon is known as hazardous polymerization and can cause fires and explosions.

Step-Growth

Step-growth polymers are defined as polymers formed by the stepwise reaction between functional groups of monomers, usually containing heteroatoms such as nitrogen or oxygen. Most step-growth polymers are also classified as condensation polymers, but not all step-growth polymers (like polyurethanes formed from isocyanate and alcohol bifunctional monomers) release condensates; in this case, we talk about addition polymers. Step-growth polymers increase in molecular weight at a very slow rate at lower conversions and reach moderately high molecular weights only at very high conversion (i.e., >95%).

To alleviate inconsistencies in these naming methods, adjusted definition for condensation and addition polymers have been developed. A condensation polymer is defined as a polymer that involves loss of small molecules during its synthesis, or contains heteroatoms as part of its backbone chain, or its repeat unit does not contain all the atoms present in the hypothetical monomer to which it can be degraded.

Chain-Growth

Chain-growth polymerization (or addition polymerization) involves the linking together of molecules incorporating double or triple carbon-carbon bonds. These unsaturated monomers (the identical molecules that make up the polymers) have extra internal bonds that are able to break and link up with other monomers to form a repeating chain, whose backbone typically contains only carbon atoms. Chain-growth polymerization is involved in the manufacture of polymers such as polyethylene, polypropylene, and polyvinyl chloride (PVC). A special case of chain-growth polymerization leads to living polymerization.

In the radical polymerization of ethylene, its π bond is broken, and the two electrons rearrange to create a new propagating center like the one that attacked it. The form this propagating center takes depends on the specific type of addition mechanism. There are several mechanisms through which this can be initiated. The free radical mechanism is one of the first methods to be used. Free radicals are very reactive atoms or molecules that have unpaired electrons. Taking the polymerization of ethylene as an example, the free radical mechanism can be divided into three stages: chain initiation, chain propagation, and chain termination.

Polymerization of ethylene.

Free radical addition polymerization of ethylene must take place at high temperatures and pressures, approximately 300 °C and 2000 atm. While most other free radical polymerizations do not require such extreme temperatures and pressures, they do tend to lack control. One effect of this lack of control is a high degree of branching. Also, as termination occurs randomly, when two chains collide, it is impossible to control the length of individual chains. A newer method of polymerization similar to free radical, but allowing more control involves the Ziegler-Natta catalyst, especially with respect to polymer branching.

Other forms of chain growth polymerization include cationic addition polymerization and anionic addition polymerization. While not used to a large extent in industry yet due to stringent reaction conditions such as lack of water and oxygen, these methods provide ways to polymerize some monomers that cannot be polymerized by free radical methods such as polypropylene. Cationic and anionic mechanisms are also more ideally suited for living polymerizations, although free radical living polymerizations have also been developed.

Esters of acrylic acid contain a carbon-carbon double bond which is conjugated to an ester group. This allows the possibility of both types of polymerization mechanism. An acrylic ester by itself can undergo chain-growth polymerization to form a homopolymer with a carbon-carbon backbone, such as poly(methyl methacrylate). Also, however, certain acrylic esters can react with diamine monomers by nucleophilic conjugate addition of amine groups to acrylic C=C bonds. In this case the polymerization proceeds by step-growth and the products are poly(beta-amino ester) copolymers, with backbones containing nitrogen (as amine) and oxygen (as ester) as well as carbon.

Physical Polymer Reaction Engineering

To produce a high-molecular-weight, uniform product, various methods are employed to better control the initiation, propagation, and termination rates during chain polymerization and also to remove excess concentrated heat during these exothermic reactions compared to polymerization of the pure monomer (also referred to as bulk polymerization). These include emulsion polymerization, solution polymerization, suspension polymerization, and precipitation polymerization. Although the polymer polydispersity and molecular weight may be improved, these methods may introduce additional processing requirements to isolate the product from a solvent.

Photopolymerization

Most photopolymerization reactions are chain-growth polymerizations which are initiated by the absorption of visible or ultraviolet light. The light may be absorbed either directly by the reactant monomer (direct photopolymerization), or else by a photosensitizer which absorbs the light and then transfers energy to the monomer. In general only the initiation step differs from that of the ordinary thermal polymerization of the same monomer; subsequent propagation, termination and chain transfer steps are unchanged. In step-growth photopolymerization, absorption of light triggers an addition (or condensation) reaction between two comonomers that do not react without light. A propagation cycle is not initiated because each growth step requires the assistance of light.

Photopolymerization can be used as a photographic or printing process, because polymerization only occurs in regions which have been exposed to light. Unreacted monomer can be removed from unexposed regions, leaving a relief polymeric image. Several forms of 3D printing—including layer-by-layer stereolithography and two-photon absorption 3D photopolymerization—use photopolymerization.

Mutliphoton polymerization using single pulses have also been demonstrated for fabrication of complex structures using a digital micromirror device.

Condensation Polymer

Condensation polymers are any kind of polymers formed through a condensation reaction—where molecules join together—losing small molecules as by-products such as water or methanol, as opposed to addition polymers which involve the reaction of unsaturated monomers. Types of condensation polymers include polyamides, polyacetals and polyesters.

Condensation polymerization, a form of step-growth polymerization, is a process by which two molecules join together, resulting in loss of small molecules which are often water. The type of end product resulting from a condensation polymerization is dependent on the number of functional end groups of the monomer which can react.

Monomers with only one reactive group terminate a growing chain, and thus give end products with a lower molecular weight. Linear polymers are created using monomers with two reactive end groups and monomers with more than two end groups give three-dimensional polymers which are crosslinked.

Dehydration synthesis often involves joining monomers with an -OH (hydroxyl) group and a freely ionized -H on either end (such as a hydrogen from the $-NH_2$ in nylon or proteins). Normally, two or more different monomers are used in the reaction. The bonds between the hydroxyl group, the hydrogen atom and their respective atoms break forming water from the hydroxyl and hydrogen, and the polymer.

Polyester is created through ester linkages between monomers, which involve the functional groups carboxyl and hydroxyl (an organic acid and an alcohol monomer).

Nylon is another common condensation polymer. It can be manufactured by reacting di-amines with carboxyl derivatives. In this example the derivative is a di-carboxylic acid, but di-acyl chlorides are also used. Another approach used is the reaction of di-functional monomers, with one amine and one carboxylic acid group on the same molecule:

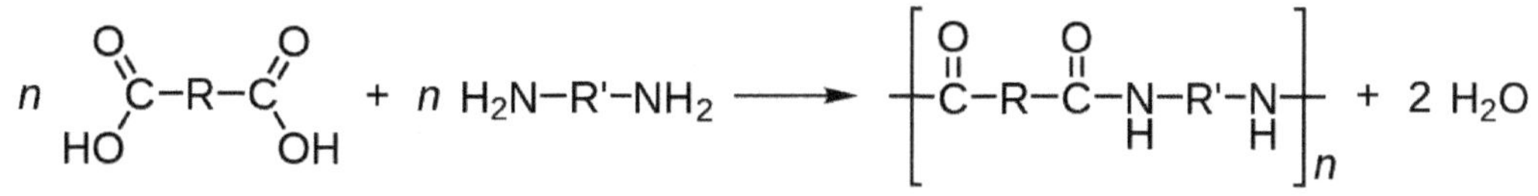

General chemical structure of one type of condensation polymer.

The carboxylic acids and amines link to form peptide bonds, also known as amide groups. Proteins are condensation polymers made from amino acid monomers. Carbohydrates are also condensation polymers made from sugar monomers such as glucose (i.e. cellulose or glycogen) and galactose.

Condensation polymerization is occasionally used to form simple hydrocarbons. This method, however, is expensive and inefficient, so the addition polymer of ethene (polyethylene) is generally used.

Condensation polymers, unlike addition polymers, may be biodegradable. The peptide or ester bonds between monomers can be hydrolysed by acid catalysts or bacterial enzymes breaking the polymer chain into smaller pieces.

The most commonly known condensation polymers are proteins, fabrics such as nylon, silk, or polyester. Characteristics of Condensation Polymers, Condensation polymers form more slowly than addition polymers, often requiring heat, and they are generally lower in molecular weight. The terminal functional groups on a chain remain active, so that groups of shorter chains combine into longer chains in the late stages of polymerization. The presence of polar functional groups on the chains often enhances chain-chain attractions, particularly if these involve hydrogen bonding, and thereby crystallinity and tensile strength. The following examples of condensation polymers are illustrative. Note that for commercial synthesis the carboxylic acid components may actually be employed in the form of derivatives such as simple esters. Also, the polymerization reactions for Nylon 6 and Spandex do not proceed by elimination of water or other small molecules. Nevertheless, the polymer clearly forms by a step-growth process.

CATALYSIS

Catalysis is the increase in the rate of a chemical reaction due to the participation of an additional substance called a catalyst. In most cases, reactions occur faster with a catalyst because they require less activation energy. Furthermore since they are not consumed in the catalyzed reaction, catalysts can continue to act repeatedly. Often only tiny amounts are required in principle.

An air filter that utilizes low-temperature oxidation catalyst used to convert carbon monoxide to less toxic carbon dioxide at room temperature. It can also remove formaldehyde from the air.

Technical Perspective

In the presence of a catalyst, less free energy is required to reach the transition state, but the total free energy from reactants to products does not change. A catalyst may participate in multiple chemical transformations. The effect of a catalyst may vary due to the presence of other substances known as inhibitors or poisons (which reduce the catalytic activity) or promoters (which increase the activity). The opposite of a catalyst, a substance that reduces the rate of a reaction, is an inhibitor.

Catalyzed reactions have a lower activation energy (rate-limiting free energy of activation) than the corresponding uncatalyzed reaction, resulting in a higher reaction rate at the same temperature and for the same reactant concentrations. However, the detailed mechanics of catalysis is complex. Catalysts may affect the reaction environment favorably, or bind to the reagents to polarize bonds, e.g. acid catalysts for reactions of carbonyl compounds, or form specific intermediates that are not produced naturally, such as osmate esters in osmium tetroxide-catalyzed dihydroxylation of alkenes, or cause dissociation of reagents to reactive forms, such as chemisorbed hydrogen in catalytic hydrogenation.

Kinetically, catalytic reactions are typical chemical reactions; i.e. the reaction rate depends on the frequency of contact of the reactants in the rate-determining step. Usually, the catalyst participates in this slowest step, and rates are limited by amount of catalyst and its “activity”. In heterogeneous catalysis, the diffusion of reagents to the surface and diffusion of products from the surface can be rate determining. A nanomaterial-based catalyst is an example of a heterogeneous catalyst. Analogous events associated with substrate binding and product dissociation apply to homogeneous catalysts.

Although catalysts are not consumed by the reaction itself, they may be inhibited, deactivated, or destroyed by secondary processes. In heterogeneous catalysis, typical secondary processes include coking where the catalyst becomes covered by polymeric side products. Additionally, heterogeneous catalysts can dissolve into the solution in a solid–liquid system or sublimate in a solid–gas system.

Background

The production of most industrially important chemicals involves catalysis. Similarly, most biochemically significant processes are catalyzed. Research into catalysis is a major field in applied science and involves many areas of chemistry, notably organometallic chemistry and materials science. Catalysis is relevant to many aspects of environmental science, e.g. the catalytic converter in automobiles and the dynamics of the ozone hole. Catalytic reactions are preferred in environmentally friendly green chemistry due to the reduced amount of waste generated, as opposed to stoichiometric reactions in which all reactants are consumed and more side products are formed. Many transition metals and transition metal complexes are used in catalysis as well. Catalysts called enzymes are important in biology.

A catalyst works by providing an alternative reaction pathway to the reaction product. The rate of the reaction is increased as this alternative route has a lower activation energy than the reaction route not mediated by the catalyst. The disproportionation of hydrogen peroxide creates water and oxygen, as shown below.

$2\ H_2O_2 \rightarrow 2\ H_2O + O_2$

This reaction is preferable in the sense that the reaction products are more stable than the starting material, though the uncatalysed reaction is slow. In fact, the decomposition of hydrogen peroxide is so slow that hydrogen peroxide solutions are commercially available. This reaction is strongly affected by catalysts such as manganese dioxide, or the enzyme peroxidase in organisms. Upon the addition of a small amount of manganese dioxide, the hydrogen peroxide reacts rapidly. This effect is readily seen by the effervescence of oxygen. The manganese dioxide is not consumed in the reaction, and thus may be recovered unchanged, and re-used indefinitely. Accordingly, manganese dioxide catalyses this reaction.

General Principles

Units

Catalytic activity is usually denoted by the symbol z and measured in mol/s, a unit which was called katal and defined the SI unit for catalytic activity since 1999. Catalytic activity is not a kind of reaction rate, but a property of the catalyst under certain conditions, in relation to a specific chemical reaction. Catalytic activity of one katal (Symbol 1 kat = 1 mol/s) of a catalyst means an amount of that catalyst (substance, in Mol) that leads to a net reaction of one Mol per second of the reactants to the resulting reagents or other outcome which was intended for this chemical reaction. A catalyst may and usually will have different catalytic activity for distinct reactions..

Typical Mechanism

Catalysts generally react with one or more reactants to form intermediates that subsequently give the final reaction product, in the process regenerating the catalyst. The following is a typical reaction scheme, where ***C*** represents the catalyst, X and Y are reactants, and Z is the product of the reaction of X and Y:

$X + \boldsymbol{C} \rightarrow X\boldsymbol{C}$ (1)

$Y + X\boldsymbol{C} \rightarrow XY\boldsymbol{C}$ (2)

$XY\boldsymbol{C} \rightarrow \boldsymbol{C}Z$ (3)

$\boldsymbol{C}Z \rightarrow \boldsymbol{C} + Z$ (4)

Although the catalyst is consumed by reaction 1, it is subsequently produced by reaction 4, so it does not occur in the overall reaction equation:

$X + Y \rightarrow Z$

As a catalyst is regenerated in a reaction, often only small amounts are needed to increase the rate of the reaction. In practice, however, catalysts are sometimes consumed in secondary processes.

The catalyst does usually appear in the rate equation. For example, if the rate-determining step in the above reaction scheme is the first step $X + C \rightarrow XC$, the catalyzed reaction will be second order with rate equation v = k_{cat}[X][C], which is proportional to the catalyst concentration [C]. However [C] remains constant during the reaction so that the catalyzed reaction is pseudo-first order: v = k_{obs}[X], where k_{obs} = k_{cat}[C].

As an example of a detailed mechanism at the microscopic level, in 2008 Danish researchers first revealed the sequence of events when oxygen and hydrogen combine on the surface of titanium dioxide (TiO_2, or titania) to produce water. With a time-lapse series of scanning tunneling microscopy images, they determined the molecules undergo adsorption, dissociation and diffusion before reacting. The intermediate reaction states were: HO_2, H_2O_2, then H_3O_2 and the final reaction product (water molecule dimers), after which the water molecule desorbs from the catalyst surface.

Reaction Energetics

Catalysts work by providing an (alternative) mechanism involving a different transition state and lower activation energy. Consequently, more molecular collisions have the energy needed to reach the transition state. Hence, catalysts can enable reactions that would otherwise be blocked or slowed by a kinetic barrier. The catalyst may increase reaction rate or selectivity, or enable the reaction at lower temperatures. This effect can be illustrated with an energy profile diagram.

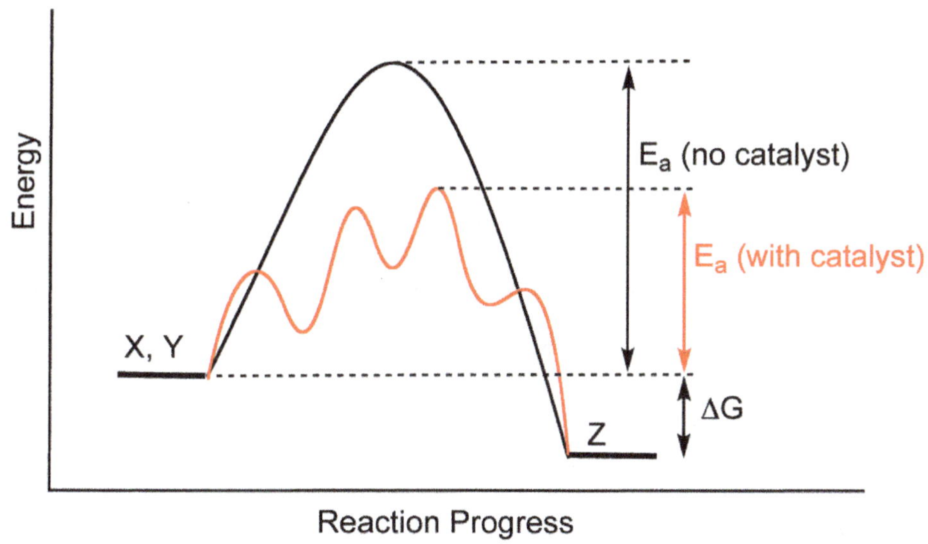

Generic potential energy diagram showing the effect of a catalyst in a hypothetical exothermic chemical reaction X + Y to give Z. The presence of the catalyst opens a different reaction pathway (shown in red) with a lower activation energy. The final result and the overall thermodynamics are the same.

In the catalyzed elementary reaction, catalysts do **not** change the extent of a reaction: they have **no** effect on the chemical equilibrium of a reaction because the rate of both the forward and the reverse reaction are both affected. The second law of thermodynamics describes why a catalyst does not change the chemical equilibrium of a reaction. Suppose there was such a catalyst that shifted an equilibrium. Introducing the catalyst to the system would result in a reaction to move to the new equilibrium, producing energy. Production of energy is a necessary result since reactions are spontaneous only if Gibbs free energy is produced, and if there is no energy barrier, there is no need for a catalyst. Then, removing the catalyst would also result in reaction, producing energy; i.e. the addition and its reverse process, removal, would both produce energy. Thus, a catalyst that could change the equilibrium would be a perpetual motion machine, a contradiction to the laws of thermodynamics.

If a catalyst does change the equilibrium, then it must be consumed as the reaction proceeds, and thus it is also a reactant. Illustrative is the base-catalysed hydrolysis of esters, where the produced carboxylic acid immediately reacts with the base catalyst and thus the reaction equilibrium is shifted towards hydrolysis.

The SI derived unit for measuring the catalytic activity of a catalyst is the katal, which is moles per second. The productivity of a catalyst can be described by the turn over number (or TON) and the catalytic activity by the turn over frequency (TOF), which is the TON per time unit. The biochemical equivalent is the enzyme unit. For more information on the efficiency of enzymatic catalysis.

The catalyst stabilizes the transition state more than it stabilizes the starting material. It decreases the kinetic barrier by decreasing the difference in energy between starting material and transition state. It does not change the energy difference between starting materials and products (thermodynamic barrier), or the available energy (this is provided by the environment as heat or light).

Materials

The chemical nature of catalysts is as diverse as catalysis itself, although some generalizations can be made. Proton acids are probably the most widely used catalysts, especially for the many reactions involving water, including hydrolysis and its reverse. Multifunctional solids often are catalytically active, e.g. zeolites, alumina, higher-order oxides, graphitic carbon, nanoparticles, nanodots, and facets of bulk materials. Transition metals are often used to catalyze redox reactions (oxidation, hydrogenation). Examples are nickel, such as Raney nickel for hydrogenation, and vanadium(V) oxide for oxidation of sulfur dioxide into sulfur trioxide by the so-called contact process. Many catalytic processes, especially those used in organic synthesis, require "late transition metals", such as palladium, platinum, gold, ruthenium, rhodium, or iridium.

Some so-called catalysts are really precatalysts. Precatalysts convert to catalysts in the reaction. For example, Wilkinson's catalyst $RhCl(PPh_3)_3$loses one triphenylphosphine ligand before entering the true catalytic cycle. Precatalysts are easier to store but are easily activated in situ. Because of this preactivation step, many catalytic reactions involve an induction period.

Chemical species that improve catalytic activity are called co-catalysts (cocatalysts) or promotors in cooperative catalysis.

Types

Catalysts can be heterogeneous or homogeneous, depending on whether a catalyst exists in the same phase as the substrate. Biocatalysts (enzymes) are often seen as a separate group.

Heterogeneous Catalysts

Heterogeneous catalysts act in a different phase than the reactants. Most heterogeneous catalysts are solids that act on substrates in a liquid or gaseous reaction mixture. Diverse mechanisms for reactions on surfaces are known, depending on how the adsorption takes place (Langmuir-Hinshelwood, Eley-Rideal, and Mars-van Krevelen). The total surface area of solid has an important effect on the reaction rate. The smaller the catalyst particle size, the larger the surface area for a given mass of particles.

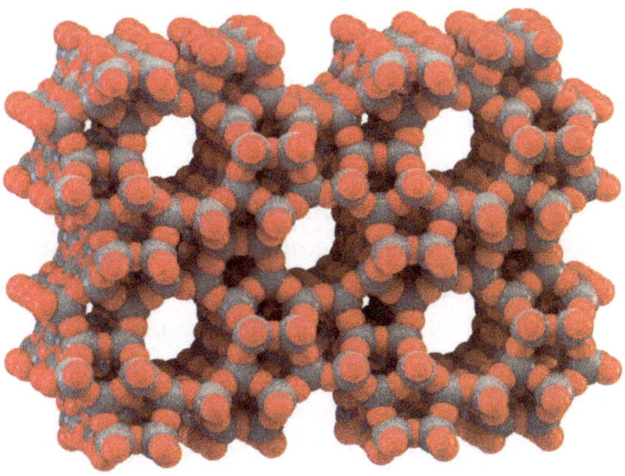

The microporous molecular structure of the zeolite ZSM-5 is exploited in catalysts used in refineries.

Zeolites are extruded as pellets for easy handling in catalytic reactors.

A heterogeneous catalyst has active sites, which are the atoms or crystal faces where the reaction actually occurs. Depending on the mechanism, the active site may be either a planar exposed metal surface, a crystal edge with imperfect metal valence or a complicated combination of the two. Thus, not only most of the volume, but also most of the surface of a heterogeneous catalyst may be catalytically inactive. Finding out the nature of the active site requires technically challenging research. Thus, empirical research for finding out new metal combinations for catalysis continues.

For example, in the Haber process, finely divided iron serves as a catalyst for the synthesis of ammonia from nitrogen and hydrogen. The reacting gases adsorb onto active sites on the iron particles. Once physically adsorbed, the reagents undergo chemisorption that results in dissociation into adsorbed atomic species, and new bonds between the resulting fragments form in part due to their close proximity. In this way the particularly strong triple bond in nitrogen is broken, which would be extremely uncommon in the gas phase due to its high activation energy. Thus, the activation energy of the overall reaction is lowered, and the rate of reaction increases. Another place where a heterogeneous catalyst is applied is in the oxidation of sulfur dioxide on vanadium(V) oxide for the production of sulfuric acid.

Heterogeneous catalysts are typically "supported," which means that the catalyst is dispersed on a second material that enhances the effectiveness or minimizes their cost. Supports prevent or reduce agglomeration and sintering of the small catalyst particles, exposing more surface area, thus catalysts have a higher specific activity (per gram) on a support. Sometimes the support is merely a surface on which the catalyst is spread to increase the surface area. More often, the support and the catalyst interact, affecting the catalytic reaction. Supports are porous materials with a high surface area, most commonly alumina, zeolites or various kinds of activated carbon. Specialized supports include silicon dioxide, titanium dioxide, calcium carbonate, and barium sulfate.

Electrocatalysts

In the context of electrochemistry, specifically in fuel cell engineering, various metal-containing catalysts are used to enhance the rates of the half reactions that comprise the fuel cell. One common type of fuel cell electrocatalyst is based upon nanoparticles of platinum that are supported on slightly larger carbon particles. When in contact with one of the electrodes in a fuel cell, this platinum increases the rate of oxygen reduction either to water, or to hydroxide or hydrogen peroxide.

Homogeneous Catalysts

Homogeneous catalysts function in the same phase as the reactants, but the mechanistic principles invoked in heterogeneous catalysis are generally applicable. Typically homogeneous catalysts are dissolved in a solvent with the substrates. One example of homogeneous catalysis involves the influence of H^+ on the esterification of carboxylic acids, such as the formation of methyl acetate from acetic acid and methanol. For inorganic chemists, homogeneous catalysis is often synonymous with organometallic catalysts.

Organocatalysis

Whereas transition metals sometimes attract most of the attention in the study of catalysis, small organic molecules without metals can also exhibit catalytic properties, as is apparent from the fact that many enzymes lack transition metals. Typically, organic catalysts require a higher loading (amount of catalyst per unit amount of reactant, expressed in mol% amount of substance) than transition metal(-ion)-based catalysts, but these catalysts are usually commercially available in bulk, helping to reduce costs. In the early 2000s, these organocatalysts were considered "new generation" and are competitive to traditional metal(-ion)-containing catalysts. Organocatalysts are supposed to operate akin to metal-free enzymes utilizing, e.g., non-covalent interactions such as hydrogen bonding. The discipline organocatalysis is divided in the application of covalent (e.g., proline, DMAP) and non-covalent (e.g., thiourea organocatalysis) organocatalysts referring to the preferred catalyst-substrate binding and interaction, respectively.

Enzymes and Biocatalysts

In biology, enzymes are protein-based catalysts in metabolism and catabolism. Most biocatalysts are enzymes, but other non-protein-based classes of biomolecules also exhibit catalytic properties including ribozymes, and synthetic deoxyribozymes.

Biocatalysts can be thought of as intermediate between homogeneous and heterogeneous catalysts, although strictly speaking soluble enzymes are homogeneous catalysts and membrane-bound enzymes are heterogeneous. Several factors affect the activity of enzymes (and other catalysts) including temperature, pH, concentration of enzyme, substrate, and products. A particularly important reagent in enzymatic reactions is water, which is the product of many bond-forming reactions and a reactant in many bond-breaking processes.

In biocatalysis, enzymes are employed to prepare many commodity chemicals including high-fructose corn syrup and acrylamide.

Some monoclonal antibodies whose binding target is a stable molecule which resembles the transition state of a chemical reaction can function as weak catalysts for that chemical reaction by lowering its activation energy. Such catalytic antibodies are sometimes called "abzymes".

Nanocatalysts

Nanocatalysts are nanomaterials with catalytic activities. They have been extensively explored for wide range of applications. Among them, the nanocatalysts with enzyme mimicking activities are collectively called as nanozymes.

Tandem Catalysis

In tandem catalysis two or more different catalysts are coupled in a one-pot reaction.

Autocatalysis

In autocatalysis the catalyst is a product of the overall reaction, in contrast to all other types of catalysis considered in this article. The simplest example of autocatalysis is a reaction of type A + B → 2 B, in one or in several steps. The overall reaction is just A → B, so that B is a product. But since B is also a reactant, it may be present in the rate equation and affect the reaction rate. As the reaction proceeds, the concentration of B increases and can accelerate the reaction as a catalyst. In effect, the reaction accelerates itself or is autocatalyzed.

A real example is the hydrolysis of an ester such as aspirin to a carboxylic acid and an alcohol. In the absence of added acid catalysts, the carboxylic acid product catalyzes the hydrolysis.

Significance

Left: Partially caramelised cube sugar, Right: burning cube sugar with ash as catalyst.

Estimates are that 90% of all commercially produced chemical products involve catalysts at some stage in the process of their manufacture. In 2005, catalytic processes generated about $900 billion in products worldwide. Catalysis is so pervasive that subareas are not readily classified. Some areas of particular concentration are surveyed below.

Energy Processing

Petroleum refining makes intensive use of catalysis for alkylation, catalytic cracking (breaking long-chain hydrocarbons into smaller pieces), naphtha reforming and steam reforming (conversion of hydrocarbons into synthesis gas). Even the exhaust from the burning of fossil fuels is treated via catalysis: Catalytic converters, typically composed of platinum and rhodium, break down some of the more harmful byproducts of automobile exhaust.

$$2\ CO + 2\ NO \rightarrow 2\ CO_2 + N_2$$

With regard to synthetic fuels, an old but still important process is the Fischer-Tropsch synthesis of hydrocarbons from synthesis gas, which itself is processed via water-gas shift reactions, catalysed by iron. Biodiesel and related biofuels require processing via both inorganic and biocatalysts.

Fuel cells rely on catalysts for both the anodic and cathodic reactions.

Catalytic heaters generate flameless heat from a supply of combustible fuel.

Bulk Chemicals

Some of the largest-scale chemicals are produced via catalytic oxidation, often using oxygen. Examples include nitric acid (from ammonia), sulfuric acid (from sulfur dioxide to sulfur trioxide by the chamber process), terephthalic acid from p-xylene, and acrylonitrile from propane and ammonia.

Many other chemical products are generated by large-scale reduction, often via hydrogenation. The largest-scale example is ammonia, which is prepared via the Haber process from nitrogen. Methanol is prepared from carbon monoxide.

Bulk polymers derived from ethylene and propylene are often prepared via Ziegler-Natta catalysis. Polyesters, polyamides, and isocyanates are derived via acid-base catalysis.

Most carbonylation processes require metal catalysts, examples include the Monsanto acetic acid process and hydroformylation.

Fine Chemicals

Many fine chemicals are prepared via catalysis; methods include those of heavy industry as well as more specialized processes that would be prohibitively expensive on a large scale. Examples include the Heck reaction, and Friedel-Crafts reactions.

Because most bioactive compounds are chiral, many pharmaceuticals are produced by enantioselective catalysis (catalytic asymmetric synthesis).

Food Processing

One of the most obvious applications of catalysis is the hydrogenation (reaction with hydrogen gas) of fats using nickel catalyst to produce margarine. Many other foodstuffs are prepared via biocatalysis (see below).

Environment

Catalysis impacts the environment by increasing the efficiency of industrial processes, but catalysis also plays a direct role in the environment. A notable example is the catalytic role of chlorine free radicals in the breakdown of ozone. These radicals are formed by the action of ultravioletradiation on chlorofluorocarbons (CFCs).

$Cl^{\cdot} + O_3 \rightarrow ClO^{\cdot} + O_2$

$ClO^{\cdot} + O^{\cdot} \rightarrow Cl^{\cdot} + O_2$

History

Generally speaking, anything that increases the rate of a process is a "catalyst", a term derived from Greek καταλύειν, meaning «to annul,» or «to untie,» or «to pick up.» The concept of catalysis was invented by chemist Elizabeth Fulhame and described in a 1794 book, based on her novel work in oxidation-reduction experiments. The term catalysis was later used by Jöns Jakob Berzelius in 1835 to describe reactions that are accelerated by substances that remain unchanged after the reaction. Fulhame, who predated Berzelius, did work with water as opposed to metals in her reduction experiments. Other 18th century chemists who worked in catalysis were Eilhard Mitscherlich who referred to it as contact processes, and Johann Wolfgang Döbereiner who spoke of contact action. He developed Döbereiner's lamp, a lighter based on hydrogen and a platinum sponge, which became a commercial success in the 1820s that lives on today. Humphry Davy discovered the use of platinum in catalysis. In the 1880s, Wilhelm Ostwald at Leipzig University started a systematic investigation into reactions that were catalyzed by the presence of acids and bases, and found that chemical reactions occur at finite rates and that these rates can be used to determine the strengths of acids and bases. For this work, Ostwald was awarded the 1909 Nobel Prize in Chemistry.

Inhibitors, Poisons, and Promoters

Substances that reduce the action of catalysts are called catalyst inhibitors if reversible, and catalyst poisons if irreversible. Promoters are substances that increase the catalytic activity, even though they are not catalysts by themselves.

Inhibitors are sometimes referred to as "negative catalysts" since they decrease the reaction rate. However the term inhibitor is preferred since they do not work by introducing a reaction path with higher activation energy; this would not reduce the rate since the reaction would continue to occur by the non-catalyzed path. Instead they act either by deactivating catalysts, or by removing reaction intermediates such as free radicals.

The inhibitor may modify selectivity in addition to rate. For instance, in the reduction of alkynes to alkenes, a palladium (Pd) catalyst partly "poisoned" with lead(II) acetate ($Pb(CH_3CO_2)_2$) can be used. Without the deactivation of the catalyst, the alkene produced would be further reduced to alkane.

The inhibitor can produce this effect by, e.g., selectively poisoning only certain types of active sites. Another mechanism is the modification of surface geometry. For instance, in hydrogenation operations, large planes of metal surface function as sites of hydrogenolysis catalysis while sites catalyzing hydrogenation of unsaturates are smaller. Thus, a poison that covers surface randomly will tend to reduce the number of uncontaminated large planes but leave proportionally more smaller sites free, thus changing the hydrogenation vs. hydrogenolysis selectivity. Many other mechanisms are also possible.

Promoters can cover up surface to prevent production of a mat of coke, or even actively remove such material (e.g., rhenium on platinum in platforming). They can aid the dispersion of the catalytic material or bind to reagents.

Current Market

The global demand on catalysts in 2010 was estimated at approximately 29.5 billion USD. With the rapid recovery in automotive and chemical industry overall, the global catalyst market is expected to experience fast growth in the next years.

REFERENCES

1. “Redox Reactions”. wiley.com.
2. Haustein, Catherine Hinga (2014). K. Lee Lerner and Brenda Wilmoth Lerner, eds. Oxidation-reduction reaction. The Gale Encyclopedia of Science. 5th edition. Farmington Hills, MI: Gale Group.
3. Bockris, John O’M.; Reddy, Amulya K. N. (1970). Modern Electrochemistry. Plenum Press. pp. 352–3.
4. Harper, Douglas. “redox”. Online Etymology Dictionary.
5. Hudlický, Miloš (1996). Reductions in Organic Chemistry. Washington, D.C.: American Chemical Society. p. 429. ISBN 0-8412-3344-6.
6. Hudlický, Miloš (1990). Oxidations in Organic Chemistry. Washington, D.C.: American Chemical Society. p. 456. ISBN 0-8412-1780-7.
7. Electrode potential values from Petrucci, R. H.; Harwood, W. S.; Herring, F. G. (2002). General Chemistry (8th ed.). Prentice-Hall. p. 832.
8. “gutier.doc” (PDF). Retrieved 2008-06-30. (2.76 MiB)
9. Robertson, William (2010). More Chemistry Basics. National Science Teachers Association. p. 82. ISBN 978-1-936137-74-9.
10. Phillips, John; Strozak, Victor; Wistrom, Cheryl (2000). Chemistry: Concepts and Applications. Glencoe McGraw-Hill. p. 558. ISBN 978-0-02-828210-7. Students often are confused when associating reduction with the gain of electrons.
11. Rodgers, Glen (2012). Descriptive Inorganic, Coordination, and Solid-State Chemistry. Brooks/Cole, Cengage Learning. p. 330. ISBN 978-0-8400-6846-0.
12. Zumdahl, Steven; Zumdahl, Susan (2009). Chemistry. Houghton Mifflin. p. 160. ISBN 978-0-547-05405-6.
13. Hudlický, Miloš (1996). Reductions in Organic Chemistry. Washington, D.C.: American Chemical Society. p. 429. ISBN 0-8412-3344-6.
14. Paul N. Rylander, “Hydrogenation and Dehydrogenation” in Ullmann’s Encyclopedia of Industrial Chemistry, Wiley-VCH, Weinheim, 2005. doi:10.1002/14356007.a13 487
15. Advanced Organic Chemistry Jerry March 2nd Edition
16. “Hydrogenation of nitrobenzene using polymer bound Ru(III) complexes as catalyst”. Ind. Jr. of Chem. Tech. 7: 280. 2000.

17. Patel, D. R. (1998). "Hydrogenation of nitrobenzene using polymer anchored Pd(II) complexes as catalyst". Journal of Molecular Catalysis. 130: 57. doi:10.1016/s1381-1169(97)00197-0.

18. C. F. H. Allen and James VanAllan (1955). "m-Toylybenzylamine". Org. Synth.; Coll. Vol., 3, p. 827

19. A. B. Mekler, S. Ramachandran, S. Swaminathan, and Melvin S. Newman (1973). "2-Methyl-1,3-Cyclohexanedione". Org. Synth. ; Coll. Vol., 5, p. 743

20. Knowles, W. S. (March 1986). "Application of organometallic catalysis to the commercial production of L-DOPA". Journal of Chemical Education. 63 (3): 222. doi:10.1021/ed063p222.

21. Atkins, Peter W. (2010). Shriver & Atkins' inorganic chemistry. (5th ed.). New York: W. H. Freeman and Co. p. 696. ISBN 978-1-4292-1820-7.

22. Blaser, Hans-Ulrich; Pugin, Benoît; Spindler, Felix; Thommen, Marc (December 2007). "From a Chiral Switch to a Ligand Portfolio for Asymmetric Catalysis". Accounts of Chemical Research. 40 (12): 1240–1250. doi:10.1021/ar7001057.

23. Mallat, T.; Orglmeister, E.; Baiker, A. (2007). "Asymmetric Catalysis at Chiral Metal Surfaces". Chemical Reviews. 107 (11): 4863–90. doi:10.1021/cr0683663. PMID 17927256.

24. H. Lindlar and R. Dubuis (1973). "Palladium Catalyst for Partial Reduction of Acetylenes". Org. Synth.; Coll. Vol., 5, p. 880

25. S. Robert E. Ireland and P. Bey (1988). "Homogeneous Catalytic Hydrogenation: Dihydrocarvone". Org. Synth.; Coll. Vol., 6, p. 459

26. Kwesi Amoa Catalytic Hydrogenation of Maleic Acid at Moderate Pressures A Laboratory Demonstration Journal of Chemical Education 2007, Vol. 84, p 1948. doi:10.1021/ed084p1948

27. van Es, T.; Staskun, B. "Aldehydes from Aromatic Nitriles: 4-Formylbenzenesulfonamide" Org. Syn., Coll. Vol. 6, p. 631 (1988). (Article)

28. Navarro, Daniela Maria do Amaral Ferraz; Navarro, Marcelo (2004). "Catalytic Hydrogenation of Organic Compounds without H2 Supply: An Electrochemical System". Journal of Chemical Education. 81 (9): 1350. doi:10.1021/ed081p1350.

29. Kubas, G. J., "Metal Dihydrogen and σ-Bond Complexes», Kluwer Academic/Plenum Publishers: New York, 2001

30. Gallezot, Pierre. «Hydrogenation - Heterogeneous» in Encyclopedia of Catalysis, Volume 4, ed. Horvath, I.T., John Wiley & Sons, 2003.

31. Horiuti, Iurô; Polanyi, M. (1934). «Exchange reactions of hydrogen on metallic catalysts». Transactions of the Faraday Society. 30: 1164. doi:10.1039/TF9343001164.

32. Johannes G. de Vries, Cornelis J. Elsevier, eds. The Handbook of Homogeneous Hydrogenation Wiley-VCH, Weinheim, 2007. ISBN 978-3-527-31161-3

33. Ian P. Freeman «Margarines and Shortenings» in Ullmann›s Encyclopedia of Industrial Chemistry, 2005, Wiley-VCH, Weinheim. doi:10.1002/14356007.a16_145

34. Maria Teresa Tarrago-Trani, Katherine M. Phillips, Linda E. Lemar, Joanne M. Holden «New and Existing Oils and Fats Used in Products with Reduced Trans-Fatty Acid Content» Journal of the American Dietetic Association 2006, Volume 106, pp. 867–880.doi:10.1016/j.jada.2006.03.010

35. «Deadly fats: why are we still eating them?». The Independent. 2008-06-10. Retrieved 2008-06-16.

36. «New York City passes trans fat ban». msnbc. 2006-12-05. Retrieved 2010-01-09.

37. «F.D.A. Gives Food Industry 3 Years to Eliminate Trans Fats». The New York Times. 2015-06-16. Retrieved 2015-06-16.

38. http://pubs.acs.org/cgi-bin/abstract.cgi/jacsat/1922/44/i06/f-pdf/f_ja01427a021.pdf

39. Schrock, Richard R.; Osborn, John A. (April 1976). «Catalytic hydrogenation using cationic rhodium complexes. I. Evolution of the catalytic system and the hydrogenation of olefins». Journal of the American Chemical Society. 98 (8): 2134–2143. doi:10.1021/ja00424a020.

40. C. Pettinari, F. Marchetti, D. Martini «Metal Complexes as Hydrogenation Catalysts» Comprehensive Coordination Chemistry II, 2004, volume 9. pp. 75–139. doi:10.1016/B0-08-043748-6/09125-8

41. Walling, Cheves.; Bollyky, Laszlo. (1964). «Homogeneous Hydrogenation in the Absence of Transition-Metal Catalysts». Journal of the American Chemical Society. 86 (18): 3750. doi:10.1021/ja01072a028.

42. Berkessel, Albrecht; Schubert, Thomas J. S.; Müller, Thomas N. (2002). «Hydrogenation without a Transition-Metal Catalyst: On the Mechanism of the Base-Catalyzed Hydrogenation of Ketones». Journal of the American Chemical Society. 124 (29): 8693–8. doi:10.1021/ja016152r. PMID 12121113.

43. Chase, Preston A.; Welch, Gregory C.; Jurca, Titel; Stephan, Douglas W. (2007). «Metal-Free Catalytic Hydrogenation». Angewandte Chemie International Edition. 46 (42): 8050. doi:10.1002/anie.200702908.

44. Li, Baojun; Xu, Zheng (2009). «A Nonmetal Catalyst for Molecular Hydrogen Activation with Comparable Catalytic Hydrogenation Capability to Noble Metal Catalyst». Journal of the American Chemical Society. 131 (45): 16380–2. doi:10.1021/ja9061097. PMID 19845383.

45. Adams, Roger; Voorhees, V. (1928). «Apparatus for catalytic reduction». Organic Syntheses. 8: 10. doi:10.15227/orgsyn.008.0010.

46. Joshi, J.B.; Pandit, A.B.; Sharma, M.M. (1982). «Mechanically agitated gas–liquid reactors». Chemical Engineering Science. 37 (6): 813. doi:10.1016/0009-2509(82)80171-1.

47. Burgess, J. (1978). Metal ions in solution. New York: Ellis Horwood.

48. Richens, D. T. (1997). The chemistry of aqua ions: synthesis, structure, and reactivity: a tour through the periodic table of the elements. Wiley. ISBN 0-471-97058-1.

49. Baes, C.F.; Mesmer, R.E. The Hydrolysis of Cations, (1976), Wiley, New York

50. Greenwood, Norman N.; Earnshaw, Alan (1997). Chemistry of the Elements (2nd ed.). Butterworth-Heinemann. p. 384. ISBN 0-08-037941-9.
51. H. Adkins; W. H. Hartung (1926). «Acrolein». Org. Synth. 6: 1. doi:10.15227/orgsyn.006.0001.
52. J. Brent Friesen; Robert Schretzman (2011). «Dehydration of 2-Methyl-1-cyclohexanol: New Findings from a Popular Undergraduate Laboratory Experiment». J. Chem. Educ. 88 (8): 1141–1147. doi:10.1021/ed900049b.
53. H. Plieninger; Gunda Keilich (1956). «Die Dienol-Benzol-Umlagerung». Angew. Chem. 68 (19): 618–618. doi:10.1002/ange.19560681914.
54. Margaret Jevnik Gentles; Jane B. Moss; Hershel L. Herzog; E. B. Hershberg (1958). «The Dienol-Benzene Rearrangement. Some Chemistry of 1,4-Androstadiene-3,17-dione». J. Am. Chem. Soc. 80 (14): 3702–3705. doi:10.1021/ja01547a058.
55. youtube clip reaction sugar with sulfuric acid
56. http://www.exo.net/~pauld/activities/astronomy/transitvenus/sugarsulfuricacid.htm
57. Yoel Sasson «Formation of Carbon–Halogen Bonds (Cl, Br, I)» in Patai›s Chemistry of Functional Groups, 2009, Wiley-VCH, Weinheim. doi:10.1002/9780470682531.pat0011
58. M. Rossberg et al. "Chlorinated Hydrocarbons" in Ullmann's Encyclopedia of Industrial Chemistry 2006, Wiley-VCH, Weinheim. doi:10.1002/14356007.a06_233.pub2
59. T. Mori; R. Rathore (1998). «X-Ray structure of bridged 2,2′-bi(adamant-2-ylidene) chloronium cation and comparison of its reactivity with a singly bonded chloroarenium cation». Chem. Commun. (8): 927–928. doi:10.1039/a709063c.
60. Illustrative procedure for chlorination of an aromatic compound: Edward R. Atkinson, Donald M. Murphy, and James E. Lufkin (1951). «dl-4,4′,6,6′-Tetrachlorodiphenic Acid». Org. Synth. ; Coll. Vol., 4, p. 872
61. Synthesis of essential drugs, Ruben Vardanyan, Victor Hruby; Elsevier 2005 ISBN 0-444-52166-6
62. Gordon W. Gribble "The diversity of naturally occurring organobromine compounds" Chemical Society Reviews, 1999, volume 28, pages 335–346. doi:10.1039/a900201d
63. Greenwood, Norman N.; Earnshaw, Alan (1997). Chemistry of the Elements (2nd ed.). Butterworth-Heinemann. ISBN 0-08-037941-9.
64. A Citizen's Guide to Chemical Dehalogenation by the U.S. EPA

CHAPTER 7

Transport Phenomena

Chapter Contents

- **Introduction**
- **Overview**
- **Commonalities Among Phenomena**
- **The Onsager Reciprocal Relations**

Transport phenomenon, in physics, any of the phenomena involving the movement of various entities, such as mass, momentum, or energy, through a medium, fluid or solid, by virtue of nonuniform conditions existing within the medium. Variations of concentration in a medium, for example, lead to the relative motion of the various chemical species present, and this mass transport is generally referred to as diffusion. Variations of velocity within a fluid result in the transport of momentum, which is normally referred to as viscous flow. Variations in temperature result in the transport of energy, a process usually called heat conduction. There are many similarities in the mathematical descriptions of these three phenomena; and the three often occur together physically, as in combustion, where a flowing, viscous, fluid mixture is undergoing chemical reactions that produce heat, which is conducted away, and that produce various chemical species that interdiffuse with one another.

INTRODUCTION

In engineering, physics and chemistry, the study of transport phenomena concerns the exchange of mass, energy, and momentum between observed and studied systems. While it draws from fields as diverse as continuum mechanics and thermodynamics, it places a heavy emphasis on the commonalities between the topics covered. Mass, momentum, and heat transport all share a very similar mathematical framework, and the parallels between them are exploited in the study of transport phenomena to draw deep mathematical connections that often provide very useful tools in the analysis of one field that are directly derived from the others.

While it draws its theoretical foundation from principles in a number of fields, most of the fundamental transport theory is a restatement of basic conservation laws.

The fundamental analyses in all three subfields of mass, hcat, and momentum transfer are often grounded in the simple principle that the sum total of the quantities being studied must be conserved by the system and its environment. Thus, the different phenomena that lead to transport are each considered individually with the knowledge that the sum of their contributions must equal zero. This principle is useful for calculating many relevant quantities. For example, in fluid mechanics, a common use of transport analysis is to determine the velocity profile of a fluid flowing through a rigid volume.

Transport phenomena are ubiquitous throughout the engineering disciplines. Some of the most common examples of transport analysis in engineering are seen in the fields of process, chemical, biological, and mechanical engineering, but the subject is a fundamental component of the curriculum in all disciplines involved in any way with fluid mechanics, heat transfer, and mass transfer. It is now considered to be a part of the engineering discipline as much as thermodynamics, mechanics, and electromagnetism.

Transport phenomena encompass all agents of physical change in the universe. Moreover, they are considered to be fundamental building blocks which developed the universe, and which is responsible for the success of all life on earth. However, the scope here is limited to the relationship of transport phenomena to artificial engineered systems.

OVERVIEW

In physics, transport phenomena are all irreversible processes of statistical nature stemming from the random continuous motion of molecules, mostly observed in fluids. Every aspect of transport phenomena is grounded in two primary concepts: the conservation laws, and the constitutive equations. The conservation laws, which in the context of transport phenomena are formulated as continuity equations, describe how the quantity being studied must be conserved. The constitutive equations describe how the quantity in question responds to various stimuli via transport. Prominent examples include Fourier's Law of Heat Conduction and the Navier-Stokes equations, which describe, respectively, the response of heat flux to temperature gradients and the relationship between fluid flux and the forces applied to the fluid. These equations also demonstrate the deep connection between transport phenomena and thermodynamics, a connection that explains why transport phenomena are irreversible. Almost all of these physical phenomena ultimately involve systems seeking their lowest energy state in keeping with the principle of minimum energy. As they approach this state, they tend to achieve true thermodynamic equilibrium, at which point there are no longer any driving forces in the system and transport ceases. The various aspects of such equilibrium are directly connected to a specific transport: heat transfer is the system's attempt to achieve thermal equilibrium with its environment, just as mass and momentum transport move the system towards chemical and mechanical equilibrium.

Examples of transport processes include heat conduction (energy transfer), fluid flow (momentum transfer), molecular diffusion (mass transfer), radiation and electric charge transfer in semiconductors.

Transport phenomena have wide application. For example, in solid state physics, the motion and interaction of electrons, holes and phonons are studied under "transport phenomena". Another example is in biomedical engineering, where some transport phenomena of interest are thermoregulation, perfusion, and microfluidics. In chemical engineering, transport phenomena are studied in reactor design, analysis of molecular or diffusive transport mechanisms, and metallurgy.

The transport of mass, energy, and momentum can be affected by the presence of external sources:

- An odor dissipates more slowly (and may intensify) when the source of the odor remains present.
- The rate of cooling of a solid that is conducting heat depends on whether a heat source is applied.
- The gravitational force acting on a rain drop counteracts the resistance or drag imparted by the surrounding air.

COMMONALITIES AMONG PHENOMENA

An important principle in the study of transport phenomena is analogy between phenomena.

Diffusion

There are some notable similarities in equations for momentum, energy, and mass transfer which can all be transported by diffusion, as illustrated by the following examples:

- Mass: the spreading and dissipation of odors in air is an example of mass diffusion.
- Energy: the conduction of heat in a solid material is an example of heat diffusion.
- Momentum: the drag experienced by a rain drop as it falls in the atmosphere is an example of momentum diffusion (the rain drop loses momentum to the surrounding air through viscous stresses and decelerates).

The molecular transfer equations of Newton's law for fluid momentum, Fourier's law for heat, and Fick's law for mass are very similar. One can convert from one transfer coefficient to another in order to compare all three different transport phenomena.

Comparison of diffusion phenomena		
Transported quantity	**Physical phenomenon**	**Equation**
Momentum	Viscosity (Newtonian fluid)	$\tau = -\nu \frac{\partial \rho v}{\partial x}$

Energy	Heat conduction (Fourier's law)	$\frac{q}{A} = -k\frac{dT}{dx}$
Mass	Molecular diffusion (Fick's law)	$J = -D\frac{\partial C}{\partial x}$

(Definitions of these formulas are given below).

A great deal of effort has been devoted in the literature to developing analogies among these three transport processes for turbulent transfer so as to allow prediction of one from any of the others. The Reynolds analogy assumes that the turbulent diffusivities are all equal and that the molecular diffusivities of momentum (μ/ρ) and mass (D_{AB}) are negligible compared to the turbulent diffusivities. When liquids are present and/or drag is present, the analogy is not valid. Other analogies, such as von Karman's and Prandtl's, usually result in poor relations.

The most successful and most widely used analogy is the Chilton and Colburn J-factor analogy. This analogy is based on experimental data for gases and liquids in both the laminar and turbulent regimes. Although it is based on experimental data, it can be shown to satisfy the exact solution derived from laminar flow over a flat plate. All of this information is used to predict transfer of mass.

Onsager Reciprocal Relations

In thermodynamics, the Onsager reciprocal relations express the equality of certain ratios between flows and forces in thermodynamic systems out of equilibrium, but where a notion of local equilibrium exists.

"Reciprocal relations" occur between different pairs of forces and flows in a variety of physical systems. For example, consider fluid systems described in terms of temperature, matter density, and pressure. In this class of systems, it is known that temperature differences lead to heat flows from the warmer to the colder parts of the system; similarly, pressure differences will lead to matter flow from high-pressure to low-pressure regions. What is remarkable is the observation that, when both pressure and temperature vary, temperature differences at constant pressure can cause matter flow (as in convection) and pressure differences at constant temperature can cause heat flow. Perhaps surprisingly, the heat flow per unit of pressure difference and the density (matter) flow per unit of temperature difference are equal. This equality was shown to be necessary by Lars Onsager using statistical mechanics as a consequence of the time reversibility of microscopic dynamics (microscopic reversibility). The theory developed by Onsager is much more general than this example and capable of treating more than two thermodynamic forces at once, with the limitation that "the principle of dynamical reversibility does not apply when (external) magnetic fields or Coriolis forces are present", in which case "the reciprocal relations break down".

Though the fluid system is perhaps described most intuitively, the high precision of electrical measurements makes experimental realizations of Onsager's reciprocity easier in systems involving electrical phenomena. In fact, Onsager's 1931 paper refers to thermoelectricity and transport phenomena in electrolytes as well-known from the 19th century, including "quasi-thermodynamic" theories by Thomson and Helmholtz respectively. Onsager's reciprocity in the thermoelectric effect manifests itself in the equality of the Peltier (heat flow caused by a voltage difference) and Seebeck (electrical current caused by a temperature difference) coefficients of a thermoelectric material. Similarly, the so-called "direct piezoelectric" (electrical current produced by mechanical stress) and "reverse piezoelectric" (deformation produced by a voltage difference) coefficients are equal. For many kinetic systems, like the Boltzmann equation or chemical kinetics, the Onsager relations are closely connected to the principle of detailed balance and follow from them in the linear approximation near equilibrium.

Experimental verifications of the Onsager reciprocal relations were collected and analyzed by D. G. Miller for many classes of irreversible processes, namely for thermoelectricity, electrokinetics, transference in electrolytic solutions, diffusion, conduction of heat and electricity in anisotropic solids, thermomagnetism and galvanomagnetism. In this classical review, chemical reactions are considered as "cases with meager" and inconclusive evidence. Further theoretical analysis and experiments support the reciprocal relations for chemical kinetics with transport.

For his discovery of these reciprocal relations, Lars Onsager was awarded the 1968 Nobel Prize in Chemistry. The presentation speech referred to the three laws of thermodynamics and then added "It can be said that Onsager's reciprocal relations represent a further law making a thermodynamic study of irreversible processes possible." Some authors have even described Onsager's relations as the "Fourth law of thermodynamics".

Example: Fluid system

The Fundamental Equation

The basic thermodynamic potential is internal energy. In a simple fluid system, neglecting the effects of viscosity the fundamental thermodynamic equation is written:

$$dU = T\,dS - P\,dV + \mu\,dM$$

where U is the internal energy, T is temperature, S is entropy, P is the hydrostatic pressure, V is the volume, μ is the chemical potential, and M mass. In terms of the internal energy density, u, entropy density s, and mass density ρ, the fundamental equation at fixed volume is written:

$$du = T\,ds + \mu\,d\rho$$

For non-fluid or more complex systems there will be a different collection of variables describing the work term, but the principle is the same. The above equation may be solved for the entropy density:

$$ds = (1/T)\,du + (-\mu/T)\,d\rho$$

The above expression of the first law in terms of entropy change defines the entropic conjugate variables of u and ρ, which are $1/T$ and $-\mu/T$ and are intensive quantities analogous to potential energies; their gradients are called thermodynamic forces as they cause flows of the corresponding extensive variables as expressed in the following equations.

The Continuity Equations

The extensive quantities U and M are conserved and their flows satisfy continuity equations. The conservation of mass is written:

$$\frac{\partial \rho}{\partial t} + \nabla \cdot \mathbf{J}_\rho = 0$$

and, assuming that fluid velocity makes a negligible contribution to the energy flow, the conservation of energy is simply the conservation of the internal energy:

$$\frac{\partial u}{\partial t} + \nabla \cdot \mathbf{J}_u = 0$$

where $\mathbf{J}_\rho$ is the mass flux vector and $\mathbf{J}_u$ is the heat flux vector.

The entropy is not conserved and its continuity equation is written:

$$\frac{\partial s}{\partial t} + \nabla \cdot \mathbf{J}_s = \frac{\partial s_c}{\partial t}$$

where $\frac{\partial s_c}{\partial t}$ is the rate of increase in entropy density due to the irreversible processes of equilibration occurring in the fluid.

The Phenomenological Equations

In the absence of matter flows, Fourier's law is usually written:

$$\mathbf{J}_u = -k\nabla T;$$

where k is the thermal conductivity. However, this law is just a linear approximation, and holds only for the case where $\nabla T \ll T$, with the thermal conductivity possibly being a function of the thermodynamic state variables, but not their gradients or time rate of change. Assuming that this is the case, Fourier's law may just as well be written:

$$\mathbf{J}_u = kT^2\nabla(1/T);$$

In the absence of heat flows, Fick's law of diffusion is usually written:

$$\mathbf{J}_\rho = -D\nabla\rho,$$

where D is the coefficient of diffusion. Since this is also a linear approximation and since the chemical potential is monotonically increasing with density at a fixed temperature, Fick's law may just as well be written:

$$\mathbf{J}_\rho = D' \nabla(-\mu/T)$$

where, again, D' is a function of thermodynamic state parameters, but not their gradients or time rate of change. For the general case in which there are both mass and energy fluxes, the phenomenological equations may be written as:

$$\mathbf{J}_u = L_{uu} \nabla(1/T) + L_{u\rho} \nabla(-\mu/T)$$
$$\mathbf{J}_\rho = L_{\rho u} \nabla(1/T) + L_{\rho\rho} \nabla(-\mu/T)$$

or, more concisely,

$$\mathbf{J}_\alpha = \sum_\beta L_{\alpha\beta} \nabla f_\beta$$

where the entropic "thermodynamic forces" conjugate to the "displacements" u and ρ are $f_u = (1/T)$ and $f_\rho = (-\mu/T)$ and $L_{\alpha\beta}$ is the Onsager matrix of phenomenological coefficients.

The Rate of Entropy Production

From the fundamental equation, it follows that:

$$\frac{\partial s}{\partial t} = (1/T)\frac{\partial u}{\partial t} + (-\mu/T)\frac{\partial \rho}{\partial t}$$

and

$$\mathbf{J}_s = (1/T)\mathbf{J}_u + (-\mu/T)\mathbf{J}_\rho = \sum_\alpha \mathbf{J}_\alpha f_\alpha$$

Using the continuity equations, the rate of entropy production may now be written:

$$\frac{\partial s_c}{\partial t} = \mathbf{J}_u \cdot \nabla(1/T) + \mathbf{J}_\rho \cdot \nabla(-\mu/T) = \sum_\alpha \mathbf{J}_\alpha \cdot \nabla f_\alpha$$

and, incorporating the phenomenological equations:

$$\frac{\partial s_c}{\partial t} = \sum_\alpha \sum_\beta L_{\alpha\beta}(\nabla f_\alpha) \cdot (\nabla f_\beta)$$

It can be seen that, since the entropy production must be greater than zero, the Onsager matrix of phenomenological coefficients $L_{\alpha\beta}$ is a positive semi-definite matrix.

THE ONSAGER RECIPROCAL RELATIONS

Onsager's contribution was to demonstrate that not only is $L_{\alpha\beta}$ positive semi-definite, it is also symmetric, except in cases where time-reversal symmetry is broken. In other words, the cross-coefficients $L_{u\rho}$ and $L_{\rho u}$ are equal. The fact that they are at least proportional follows from simple dimensional analysis (*i.e.*, both coefficients are measured in the same units of temperature times mass density).

The rate of entropy production for the above simple example uses only two entropic forces, and a 2x2 Onsager phenomenological matrix. The expression for the linear approximation to the fluxes and the rate of entropy production can very often be expressed in an analogous way for many more general and complicated systems.

Abstract Formulation

Let $x_1, x_2, \ldots, x_n$ denote fluctuations from equilibrium values in several thermodynamic quantities, and let $S(x_1, x_2, \ldots, x_n)$ be the entropy. Then, Boltzmann's entropy formula gives for the probability distribution function $w = A\exp(S/k)$, *A*=const, since the probability of a given set of fluctuations $x_1, x_2, \ldots, x_n$ is proportional to the number of microstates with that fluctuation. Assuming the fluctuations are small, the probability distribution function can be expressed through the second differential of the entropy

$$w = \tilde{A}e^{-\frac{1}{2}\beta_{ik}x_i x_k}; \qquad \beta_{ik} = -\frac{1}{k}\frac{\partial^2 S}{\partial x_i \partial x_k},$$

where we are using Einstein summation convention and β_{ik} is a positive definite symmetric matrix. Using the quasi-stationary equilibrium approximation, that is, assuming that the system is only slightly non-equilibrium, we have $\dot{x}_i = -\lambda_{ik}x_k$

Suppose we define *thermodynamic conjugate* quantities as $X_i = -\frac{1}{k}\frac{\partial S}{\partial x_i}$, which can also be expressed as linear functions (for small fluctuations): $X_i = \beta_{ik}x_k$

Thus, we can write $\dot{x}_i = -\gamma_{ik}X_k$ where $\gamma_{ik} = \lambda_{il}\beta^{-1}_{lk}$ are called *kinetic coefficients*

The *principle of symmetry of kinetic coefficients* or the *Onsager's principle* states that γ is a symmetric matrix, that is $\gamma_{ik} = \gamma_{ki}$

Proof

Define mean values $\xi_i(t)$ and $\Xi_i(t)$ of fluctuating quantities x_i and X_i respectively such that they take given values $x_1, x_2, \ldots$ at $t = 0$ Note that $\dot{\xi}_i(t) = -\gamma_{ik}\Xi_k$

Symmetry of fluctuations under time reversal implies that

$$\langle x_i(t)x_k(0)\rangle = \langle x_i(-t)x_k(0)\rangle = \langle x_i(0)x_k(t)\rangle$$

or, with $\xi_i(t)$, we have $\langle \xi_i(t)x_k\rangle = \langle x_i\xi_k(t)\rangle$

Differentiating with respect to t and substituting, we get $\gamma_{il}\langle \Xi_l(t)x_k\rangle = \gamma_{kl}\langle x_i\Xi_l(t)\rangle$

Putting $t = 0$ in the above equation, $\gamma_{il}\langle X_l x_k\rangle = \gamma_{kl}\langle X_l x_i\rangle$

It can be easily shown from the definition that $\langle X_i x_k\rangle = \delta_{ik}$, and hence, we have the required result.

Momentum Transfer

In momentum transfer, the fluid is treated as a continuous distribution of matter. The study of momentum transfer, or fluid mechanics can be divided into two branches: fluid statics (fluids at rest), and fluid dynamics (fluids in motion). When a fluid is flowing in the x direction parallel to a solid surface, the fluid has x-directed momentum, and its concentration is $_x\rho$. By random diffusion of molecules there is an exchange of molecules in the z direction. Hence the x-directed momentum has been transferred in the z-direction from the faster- to the slower-moving layer. The equation for momentum transport is Newton's Law of Viscosity written as follows:

$$\tau_{zx} = -\nu\frac{\partial \rho v_x}{\partial z}$$

where τ_{zx} is the flux of x-directed momentum in the z direction, is $\mu/$, the momentum diffusivity, z is the distance of transport or diffusion, is the density, and μ is the viscosity. Newtons Law is the simplest relationship between the flux of momentum and the velocity gradient.

Mass Transfer

When a system contains two or more components whose concentration vary from point to point, there is a natural tendency for mass to be transferred, minimizing any concentration difference within the system. Mass Transfer in a system is governed by Fick's First Law: 'Diffusion flux from higher concentration to lower concentration is proportional to the gradient of the concentration of the substance and the diffusivity of the substance in the medium.' Mass transfer can take place due to different driving forces. Some of them are:

- Mass can be transferred by the action of a pressure gradient(pressure diffusion)
- Forced diffusion occurs because of the action of some external force
- Diffusion can be caused by temperature gradients (thermal diffusion)
- Diffusion can be caused by differences in chemical potential

This can be compared to Fick's Law of Diffusion:

$$J_{Ay} = -D_{AB}\frac{\partial Ca}{\partial y}$$

where D is the diffusivity constant.

Energy Transfer

All processes in engineering involve the transfer of energy. Some examples are the heating and cooling of process streams, phase changes, distillations, etc. The basic principle is the first law of thermodynamics which is expressed as follows for a static system:

$$q = -k\frac{dT}{dx}$$

The net flux of energy through a system equals the conductivity times the rate of change of temperature with respect to position.

For other systems that involve either turbulent flow, complex geometries or difficult boundary conditions another equation would be easier to use:

$$Q = h \cdot A \cdot \Delta T$$

where A is the surface area, : ΔT is the temperature driving force, Q is the heat flow per unit time, and h is the heat transfer coefficient.

Within heat transfer, two types of convection can occur:

Forced convection can occur in both laminar and turbulent flow. In the situation of laminar flow in circular tubes, several dimensionless numbers are used such as Nusselt number, Reynolds number, and Prandtl. The commonly used equation is:

$$Nu_a = \frac{h_a D}{k}$$

Natural or free convection is a function of Grashof and Prandtl numbers. The complexities of free convection heat transfer make it necessary to mainly use empirical relations from experimental data.

Heat transfer is analyzed in packed beds, reactors and heat exchangers.

REFERENCES

1. Truskey, George; Yuan F; Katz D. Transport Phenomena in Biological Systems (Second ed.). Prentice Hall. p. 888. ISBN 978-0131569881.
2. Plawsky, Joel L. (April 2001). Transport phenomena fundamentals (Chemical Industries Series). CRC Press. pp. 1, 2, 3. ISBN 978-0-8247-0500-8.
3. Plawsky, Joel., “Transport Phenomena Fundamentals.” Marcel Dekker Inc.,2009
4. Alonso & Finn. “Physics.” Addison Wesley,1992. Chapter 18
5. Deen, William M. “Analysis of Transport Phenomena.” Oxford University Press. 1998
6. J. M. Ziman, Electrons and Phonons: The Theory of Transport Phenomena in Solids (Oxford Classic Texts in the Physical Sciences)
7. Welty, James R.; Wicks, Charles E.; Wilson, Robert Elliott (1976). Fundamentals of momentum, heat, and mass transfer (2 ed.). Wiley.

8. “Thomas, William J. “Introduction to Transport Phenomena.” Prentice Hall: Upper Saddle River, NJ, 2000.
9. Transport Phenomena (1 ed.). Nirali Prakashan. 2006. p. 15-3. ISBN 81-85790-86-8., Chapter 15, p. 15-3
10. Reciprocal Relations in Irreversible Processes. I., Phys. Rev. 37, 405 - 426 (1931)
11. “Griskey, Richard G. “Transport Phenomena and Unit Operations.” Wiley & Sons: Hoboken, 2006. 228-248.
12. L. Onsager, Reciprocal Relations in Irreversible Processes. I., Phys. Rev. 37, 405–426 (1931)
13. D. G. Miller, Thermodynamics of irreversible processes. The experimental verification of the Onsager reciprocal relations, Chem. Rev. 60 (1960), 15–37.
14. G. S. Yablonsky, A. N. Gorban, D. Constales, V. V. Galvita and G. B. Marin, Reciprocal relations between kinetic curves, EPL, 93 (2011) 20004.
15. The Nobel Prize in Chemistry 1968. Presentation Speech.
16. For example Richard P. Wendt, Journal of Chemical Education v.51, p.646 (1974) “Sîmplified Transport Theory for Electrolyte Solutions”
17. Landau, L. D.; Lifshitz, E.M. (1975). Statistical Physics, Part 1. Oxford, UK: Butterworth-Heinemann. ISBN 978-81-8147-790-3.

CHAPTER 8

Chemical Plant Operations

Chapter Contents

A chemical plant is an industrial facility designed for the purpose of processing and producing chemicals. Chemical plants can be found all over the world, producing a wide range of products, from photographic chemicals used in developing rooms to herbicides sprayed on commercially produced crops. Numerous specialty facilities such as wastewater treatment plants operate using methods similar to those seen at chemical plants.

At a chemical plant, the goal is to produce new chemicals, or to process chemical products. A variety of processing techniques are used, depending on what the plant manufactures, and most chemical plants make many different types of chemicals which are related to each other. A common feature of chemical plants is large vats or tanks where chemicals are processed and stored.

INTRODUCTION

A chemical plant is an industrial process plant that manufactures (or otherwise processes) chemicals, usually on a large scale. The general objective of a chemical plant is to create new material wealth via the chemical or biological transformation and or separation of materials. Chemical plants use specialized equipment, units, and technology in the manufacturing process. Other kinds of plants, such as polymer, pharmaceutical, food, and some beverage production facilities, power plants, oil refineries or other refineries, natural gas processing and biochemical plants, water and wastewater treatment, and pollution control equipment use many technologies that have similarities to chemical plant technology such as fluid systems and chemical reactor systems. Some would consider an oil refinery or a pharmaceutical or polymer manufacturer to be effectively a chemical plant.

BASF Chemical Plant Portsmouth Site in the West Norfolk area of Portsmouth, Virginia, United States. The plant is served by the Commonwealth Railway.

Petrochemical plants (plants using chemicals from petroleum as a raw material or *feedstock*) are usually located adjacent to an oil refinery to minimize transportation costs for the feedstocks produced by the refinery. Speciality chemical and fine chemical plants are usually much smaller and not as sensitive to location. Tools have been developed for converting a base project cost from one geographic location to another.

Chemical Processes

Chemical plants use chemical processes, which are detailed industrial-scale methods, to transform feedstock chemicals into products. The same chemical process can be used at more than one chemical plant, with possibly differently scaled capacities at each plant. Also, a chemical plant at a site may be constructed to utilize more than one chemical process, for instance to produce multiple products.

A chemical plant commonly has usually large vessels or sections called units or lines that are interconnected by piping or other material-moving equipment which can carry streams of material. Such material streams can include fluids (gas or liquid carried in piping) or sometimes solids or mixtures such as slurries. An overall chemical process is commonly made up of steps called unit operations which occur in the individual units. A raw material going into a chemical process or plant as input to be converted into a product is commonly called a feedstock, or simply feed. In addition to feedstocks for the plant as a whole, an input stream of material to be processed in a particular unit can similarly be considered feed for that unit. Output streams from the plant as a whole are final products and sometimes output streams from individual units may be considered intermediate products for their units. However, final products from one plant may be intermediate chemicals used as feedstock in another plant for further processing. For example, some products from an oil refinery may used as feedstock in petrochemical plants, which may in turn produce feedstocks for pharmaceutical plants.

Either the feedstock(s), the product(s), or both may be individual compounds or mixtures. It is often not worthwhile separating the components in these mixtures completely; specific levels of purity depend on product requirements and process economics.

Operations

Chemical processes may be run in continuous or batch operation.

Batch Operation

In batch operation, production occurs in time-sequential steps in discrete batches. A batch of feedstock(s) is fed (or charged) into a process or unit, then the chemical process takes place, then the product(s) and any other outputs are removed. Such batch production may be repeated over again and again with new batches of feedstock. Batch operation is commonly used in smaller scale plants such as pharmaceutical or specialty chemicals production, for purposes of improved traceability as well as flexibility. Continuous plants are usually used to manufacture commodity or petrochemicals while batch plants are more common in speciality and fine chemical production as well as pharmaceutical active ingredient (API) manufacture.

Continuous Operation

In continuous operation, all steps are ongoing continuously in time. During usual continuous operation, the feeding and product removal are ongoing streams of moving material, which together with the process itself, all take place simultaneously and continuously. Chemical plants or units in continuous operation are usually in a steady state or approximate steady state. Steady state means that quantities related to the process do not change as time passes during operation. Such constant quantities include stream flow rates, heating or cooling rates, temperatures, pressures, and chemical compositions at any given point (location). Continuous operation is more efficient in many large scale operations like petroleum refineries. It is possible for some units to operate continuously and others be in batch operation in a chemical plant. The amount of primary feedstock or product per unit of time which a plant or unit can process is referred to as the capacity of that plant or unit. For examples: the capacity of an oil refinery may be given in terms of barrels of crude oil refined per day; alternatively chemical plant capacity may be given in tons of product produced per day. In actual daily operation, a plant (or unit) will operate at a percentage of its full capacity. Engineers typically assume 90% operating time for plants which work primarily with fluids, and 80% uptime for plants which primarily work with solids.

UNITS AND FLUID SYSTEMS

Specific unit operations are conducted in specific kinds of units. Although some units may operate at ambient temperature or pressure, many units operate at higher or lower temperatures or pressures. Vessels in chemical plants are often cylindrical with rounded ends, a shape which can be suited to hold either high pressure or vacuum. Chemical reactions can convert certain kinds of compounds into other compounds in chemical reactors. Chemical reactors may be packed beds and may have solid heterogeneous catalysts which stay in the reactors as fluids move through, or may be simply be stirred vessels in which reactions occur. Since the surface of solid heterogeneous catalysts may sometimes become "poisoned" from deposits such as coke, regeneration of catalysts may be necessary. Fluidized beds may also be used in some cases to ensure good mixing. There can also be units (or subunits) for mixing (including dissolving), separation, heating, cooling, or some combination of these. For example, chemical reactors often have stirring for mixing and heating or cooling to maintain temperature. When designing plants on a large scale, heat produced or absorbed by chemical reactions must be considered. Some plants may have units with organism cultures for biochemical processes such as fermentation or enzyme production.

Distillation unit in Italy

Separation processes include filtration, settling (sedimentation), extraction or leaching, distillation, recrystallization or precipitation (followed by filtration or settling), reverse osmosis, drying, and adsorption. Heat exchangers are often used for heating or cooling, including boiling or condensation, often in conjunction with other units such as distillation towers. There may also be storage tanks for storing feedstock, intermediate or final products, or waste. Storage tanks commonly have level indicators to show how full they are. There may be structures holding or supporting sometimes massive units and their associated equipment. There are often stairs, ladders, or other steps for personnel to reach points in the units for sampling, inspection, or maintenance. An area of a plant or facility with numerous storage tanks is sometimes called a *tank farm*, especially at an oil depot.

Fluid systems for carrying liquids and gases include piping and tubing of various diameter sizes, various types of valves for controlling or stopping flow, pumps for moving or pressurizing liquid, and compressors for pressurizing or moving gases. Vessels, piping, tubing, and sometimes other equipment at high or very low temperature are commonly covered with insulation for personnel safety and to maintain temperature inside. Fluid systems and units commonly have instrumentation such as temperature and pressure sensors and flow measuring devices at select locations in a plant. Online analyzers for chemical or physical property analysis have become more common. Solvents can sometimes be used to dissolve reactants or materials such as solids for extraction or leaching, to provide a suitable medium for certain chemical reactions to run, or so they can otherwise be treated as fluids.

CHEMICAL PLANT DESIGN

Today, the fundamental aspects of designing chemical plants are done by chemical engineers. Historically, this was not always the case and many chemical plants were constructed in a haphazard way before the discipline of chemical engineering became established. Chemical engineering was first established as a profession in the United Kingdom when the first chemical engineering course was given at the University of Manchester in 1887 by George E. Davis in the form of twelve lectures covering various aspects of industrial chemical practice. As a consequence George E. Davis is regarded as the World's first Chemical Engineer. Today Chemical Engineering is a profession and those Professional Chemical Engineers with experience can gain "Chartered" engineer status through the Institution of Chemical Engineers.

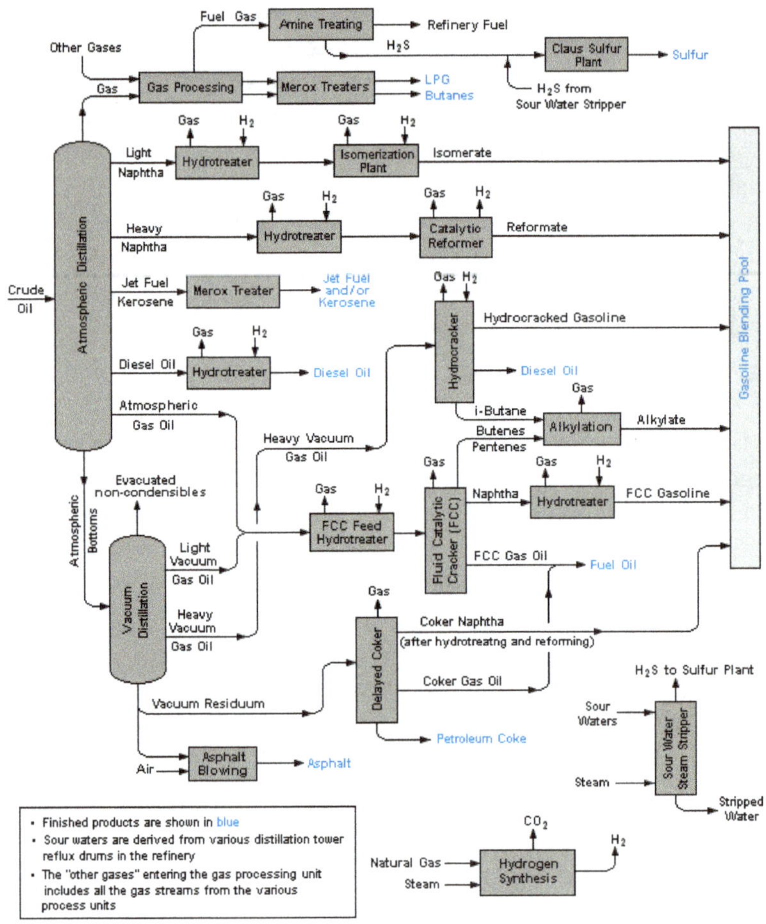

Flow diagram for a typical oil refinery

In plant design, typically less than 1 per cent of ideas for new designs ever become commercialized. During this solution process, typically, cost studies are used as an initial screening to eliminate unprofitable designs. If a process appears profitable, then other factors are considered, such as safety, environmental constraints, controllability, etc. The general goal in plant design, is to construct or synthesize "optimum designs" in the neighborhood of the desired constraints.

Many times chemists research chemical reactions or other chemical principles in a laboratory, commonly on a small scale in a "batch-type" experiment. Chemistry information obtained is then used by chemical engineers, along with expertise of their own, to convert to a chemical process and scale up the batch size or capacity. Commonly, a small chemical plant called a pilot plant is built to provide design and operating information before construction of a large plant. From data and operating experience obtained from the pilot plant, a scaled-up plant can be designed for higher or full capacity. After the fundamental aspects of a plant design are determined, mechanical or electrical engineers may become involved with mechanical or electrical details, respectively. Structural engineers may become involved in the plant design to ensure the structures can support the weight of the units, piping, and other equipment.

The units, streams, and fluid systems of chemical plants or processes can be represented by block flow diagrams which are very simplified diagrams, or process flow diagrams which are somewhat more detailed. The streams and other piping are shown as lines with arrow heads showing usual direction of material flow. In block diagrams, units are often simply shown as blocks. Process flow diagrams may use more detailed symbols and show pumps, compressors, and major valves. Likely values or ranges of material flow rates for the various streams are determined based on desired plant capacity using material balance calculations. Energy balances are also done based on heats of reaction, heat capacities, expected temperatures and pressures at various points to calculate amounts of heating and cooling needed in various places and to size heat exchangers. Chemical plant design can be shown in fuller detail in a piping and instrumentation diagram (P&ID) which shows all piping, tubing, valves, and instrumentation, typically with special symbols. Showing a full plant is often complicated in a P&ID, so often only individual units or specific fluid systems are shown in a single P&ID.

In the plant design, the units are sized for the maximum capacity each may have to handle. Similarly, sizes for pipes, pumps, compressors, and associated equipment are chosen for the flow capacity they have to handle. Utility systems such as electric power and water supply should also be included in the plant design. Additional piping lines for non-routine or alternate operating procedures, such as plant or unit startups and shutdowns, may have to be included. Fluid systems design commonly includes isolation valves around various units or parts of a plant so that a section of a plant could be isolated in case of a problem such as a leak in a unit. If pneumatically or hydraulically actuated valves are used, a system of pressurizing lines to the actuators is needed. Any points where process samples may have to be taken should have sampling lines, valves, and access to them included in the detailed design. If necessary, provisions should be made for reducing high pressure or temperature of a sampling stream, such including a pressure reducing valve or sample cooler.

Units and fluid systems in the plant including all vessels, piping, tubing, valves, pumps, compressors, and other equipment must be rated or designed to be able to withstand the entire range of pressures, temperatures, and other conditions which they could possibly encounter, including any appropriate safety factors. All such units and equipment should also be checked for materials compatibility to ensure they can withstand long-term exposure to the chemicals they will come in contact with. Any closed system in a plant which has a means of pressurizing possibly beyond the rating of its equipment, such as heating, exothermic reactions, or certain pumps or compressors, should have an appropriately sized pressure relief valve included to prevent overpressurization for safety. Frequently all of these parameters (temperatures, pressures, flow, etc.) are exhaustively analyzed in combination through a *Hazop* or *fault tree analysis*, to ensure that the plant has no known risk of serious hazard.

Within any constraints the plant is subject to, design parameters are optimized for good economic performance while ensuring safety and welfare of personnel and the surrounding community. For flexibility, a plant may be designed to operate in a range around some optimal design parameters in case feedstock or economic conditions change and re-optimization is desirable. In more modern times, computer simulations or other computer calculations have been used to help in chemical plant design or optimization.

PLANT OPERATION

Process Control

In process control, information gathered automatically from various sensors or other devices in the plant is used to control various equipment for running the plant, thereby controlling operation of the plant. Instruments receiving such information signals and sending out control signals to perform this function automatically are process *controllers*. Previously, pneumatic controls were sometimes used. Electrical controls are now common. A plant often has a control room with displays of parameters such as key temperatures, pressures, fluid flow rates and levels, operating positions of key valves, pumps and other equipment, etc. In addition, operators in the control room can control various aspects of the plant operation, often including overriding automatic control. Process control with a computer represents more modern technology. Based on possible changing feedstock composition, changing products requirements or economics, or other changes in constraints, operating conditions may be re-optimized to maximize profit.

Workers

As in any industrial setting, there are a variety of workers working throughout a chemical plant facility, often organized into departments, sections, or other work groups. Such workers typically include engineers, plant operators, and maintenance technicians. Other personnel at the site could include chemists, management/administration and office workers.

Workers in Italy, 1969. Photo by Paolo Monti

Types of engineers involved in operations or maintenance may include chemical process engineers, mechanical engineers for maintaining mechanical equipment, and electrical/computer engineers for electrical or computer equipment.

Transport

Large quantities of fluid feedstock or product may enter or leave a plant by pipeline, railroad tank car, or tanker truck. For example, petroleum commonly comes to a refinery by pipeline. Pipelines can also carry petrochemical feedstock from a refinery to a nearby petrochemical plant. Natural gas is a product which comes all the way from a natural gas processing plant to final consumers by pipeline or tubing. Large quantities of liquid feedstock are typically pumped into process units. Smaller quantities of feedstock or product may be shipped to or from a plant in drums. Use of drums about 55 gallons in capacity is common for packaging industrial quantities of chemicals. Smaller batches of feedstock may be added from drums or other containers to process units by workers.

Maintenance

In addition to feeding and operating the plant, and packaging or preparing the product for shipping, plant workers are needed for taking samples for routine and troubleshooting analysis and for performing routine and non-routine maintenance. Routine maintenance can include periodic inspections and replacement of worn catalyst, analyzer reagents, various sensors, or mechanical parts. Non-routine maintenance can include investigating problems and then fixing them, such as leaks, failure to meet feed or product specifications, mechanical failures of valves, pumps, compressors, sensors, etc.

Statutory and Regulatory Compliance

When working with chemicals, safety is a concern in order to avoid problems such as chemical accidents . In the United States, the law requires that employers provide workers working with chemicals with access to a Material Safety Data Sheet (MSDS) for every kind of chemical they work with. An MSDS for a certain chemical is prepared and provided by the supplier to whoever buys the chemical. Other laws covering chemical safety, hazardous waste, and pollution must be observed, including statutes such as the Resource Conservation and Recovery Act (RCRA) and the Toxic Substances Control Act (TSCA), and regulations such as the Chemical Facility Anti-Terrorism Standards in the United States. Hazmat (hazardous materials) teams are trained to deal with chemical leaks or spills. Process Hazard Analysis (PHA) is used to assess potential hazards in chemical plants. In 1998, the U. S. Chemical Safety and Hazard Investigation Board has become operational.

PLANT FACILITIES

The actual production or process part of a plant may be indoors, outdoors, or a combination of the two. It may be a traditional stick-built plant or a modular skid. Large modular skids are especially impressive feats of engineering. A modular skid is built including all of the modular equipment needed to do the same job a traditional stick-build plant may perform. However, the modular skid is built within a structural steel frame, allowing it to be shipped to the onsite location without needing to be rebuilt onsite. A modular skid build results in a higher functioning end product, as less hands are required in the onsite setup of the modular skid process unit, resulting in minimized risk for mishaps. The actual production section of a facility usually has the appearance of a rather industrial environment. Hard hats and work shoes are commonly worn. Floors and stairs are often made of metal grating, and there is practically no decoration. There may also be pollution control or waste treatment facilities or equipment. Sometimes existing plants may be expanded or modified based on changing economics, feedstock, or product needs. As in other production facilities, there may be shipping and receiving, and storage facilities. In addition, there are usually certain other facilities, typically indoors, to support production at the site.

Although some simple sample analysis may be able to be done by operations technicians in the plant area, a chemical plant typically has a laboratory where chemists analyze samples taken from the plant. Such analysis can include chemical analysis or determination of physical properties. Sample analysis can include routine quality control on feedstock coming into the plant, intermediate and final products to ensure quality specifications are met. Non-routine samples may be taken and analyzed for investigating plant process problems also. A larger chemical company often has a research laboratory for developing and testing products and processes where there may be pilot plants, but such a laboratory may be located at a site separate from the production plants.

A plant may also have a workshop or maintenance facility for repairs or keeping maintenance equipment. There is also typically some office space for engineers, management or administration, and perhaps for receiving visitors. The decorum there is commonly more typical of an office environment.

Clustering of Commodity Chemical Plants

Chemical Plants used particularly for commodity chemical and petrochemical manufacture, are located in relatively few manufacturing locations around the world largely due to infrastructural needs. This is less important for speciality or fine chemical batch plants. Not all commodity/petrochemicals are produced in any one location but groups of related materials often are, to induce industrial symbiosis as well as material, energy and utility efficiency and other economies of scale. These manufacturing locations often have business clusters of units called chemical plants that share utilities and large scale infrastructure such as power stations, port facilities, road and rail terminals. In the United Kingdom for example there are four main locations for commodity chemical manufacture: near the River Mersey in Northwest England, on the Humber on the East coast of Yorkshire, in Grangemouth near the Firth of Forth in Scotland and on Teesside as part of the Northeast of England Process Industry Cluster (NEPIC). Approximately 50% of the UK's petrochemicals, which are also commodity chemicals, are produced by the industry cluster companies on Teesside at the mouth of the River Tees on three large chemical parks at Wilton, Billingham and Seal Sands.

Corrosion and use of new materials

Corrosion in chemical process plants is a major issue that consumes billions of dollars yearly. Electrochemical corrosion of metals is pronounced in chemical process plants due to the presence of acid fumes and other electrolytic interactions. Recently, FRP (Fibre-reinforced plastic) is used as a material of construction. The British standard specification BS4994 is widely used for design and construction of the vessels, tanks, etc.

Chemical Accident

A chemical accident is the unintentional release of one or more hazardous substances which could harm human health or the environment. Chemical hazards are systems where chemical accidents could occur under certain circumstances. Such events include fires, explosions, leakages or releases of toxic or hazardous materials that can cause people illness, injury, disability or death.

While chemical accidents may occur whenever toxic materials are stored, transported or used, the most severe accidents are industrial accidents, involving major chemical manufacturing and storage facilities. The most significant chemical accident in recorded history was the 1984 Bhopal disaster in India, in which more than 3,000 people were killed after a highly toxic vapour, (methyl isocyanate), was released at a Union CarbidePesticides factory.

Efforts to prevent accidents range from improved safety systems to fundamental changes in chemical use and manufacture, referred to as primary prevention or inherent safety.

In the United States, concern about chemical accidents after the Bhopal disaster led to the passage of the 1986 Emergency Planning and Community Right-to-Know Act. The EPCRA requires local emergency planning efforts throughout the country, including emergency notifications. The law also requires companies to make publicly available information about their storage of toxic chemicals. Based on such information, citizens can identify the vulnerable zones in which severe toxic releases could cause harm or death.

In 1990, the Chemical Safety and Hazard Investigation Board was established by Congress, though the CSB did not become operational until 1998. The Board's mission is to determine the root causes of chemical accidents and issue safety recommendations to prevent future Safety Performance Indicators. It also organizes workshops on a number of issues related to preparing for, preventing, and responding to chemical accidents.

In the European Union, incidents such as the Flixborough disaster and the Seveso disaster led to legislation such as the Seveso Directive and Seveso planning and provide for safety reports to local authorities. Many countries have organisations that can assist with substance risk assessment and emergency planning that is required by a wide variety of legislation, such as the National Chemical Emergency Centre in the UK, Brandweerinformatiecentrum voor gevaarlijke stoffen/Fire service information centre for dangerous goods in Belgium.

In the UK, the UK Chemical Reaction Hazards Forum publishes reports of accidents on its web site. These accidents were, at the time, minor in nature, but they could have escalated into major accidents. It is hoped that publishing these incidents will prevent "Re-inventing the Wheel". At present, (Dec 2008), there are over 140 articles on the web site.

CHEMICAL PLANT COST INDEXES

Chemical plant cost indexes are dimensionless numbers employed to updating capital cost required to erect a chemical plant from a past date to a later time, following changes in the value of money due to inflation and deflation. Since, at any given time, the number of chemical plants is insufficient to use in a preliminary or predesign estimate, cost indexes are handy for a series of management purposes, like long-range planning, budgeting and escalating or de-escalating contract costs.

A cost index is the ratio of the actual price in a time period compared to that in a selected base period (a defined point in time or the average price in a certain year), multiplied by 100. Raw materials, products and energy prices, labor and construction costs change at different rates, and plant construction cost indexes are actually a composite, able to compare generic chemical plants capital costs.

Calculations Methods

Although the available indexes are compiled in several ways, they are defined to encompass a specific set of conditions and items. Prices for these items can be obtained either through company's purchasing department data or through published indexes, like those published by the Bureau of Labor Statistics (BLS). Both the CE and the IC indexes, from Chemical Engineering and Intratec, respectively, employs BLS' data.

The weighting of such factors, which is defined by the realized degree of importance of each component in the specific index (or sub-index), aims to represent the costs variations typically found for chemical plants.

Using a Cost Index

To update an item cost (equipment, projects) from period A to period B, is necessary to multiply period A's cost by the ratio of period B's index over period A's index, according to the following equation:

$$\text{Cost at B} = \text{Cost at A} \cdot \frac{\text{index at B}}{\text{index at A}}$$

As a rule-of-thumb, cost indexes permit fairly accurate estimates for cost escalation if the difference between period A and period B is less than 10 years. Differences between the actual equipment and labor prices and those predicted by the index tend to grow over the years, surpassing the typical error verified in budget-level estimates.

Cost Indexes in Engineering

The selection of the proper index to use depends on the industry in which it is applied. For example, while CE, M&S or IC Index are typically employed for chemical process industries, the ENR (Engineering News-Record) construction index is used for general industrial construction and takes in account the prices for fixed amounts of structural steel, cement, lumber and labor.

The majority of cost indexes demonstrate a time lag, due to data collection and its compilation for publishing. As stated before, some indexes use information published by other organizations and a delay in data may be verified (like those provide by the BLS). Exceptions to this are the ENR construction and the IC indexes, which present relatively current values.

The main indexes available for process industries include:

- Chemical Engineering Index, CE: composed of 4 major components – equipment, construction labor, buildings, and engineering and supervision – the index is employed primary as a process plant construction index, was established using a base period of 1957-1959 as 100. The CE Index is updated monthly and it lags in time by about 3 months. The CE Index was revised in 1982, to account for changes in labor productivity and, again, in 2002. Published in each issue of Chemical Engineering.
- Marshall and Swift Cost Index, M&S (originally known as Marshall and Stevens Index): a composite of two major components - process-industry equipment average and all-industry equipment average - was established in 1926 with a value of 100. Some industries considered in the process-industry equipment average are chemicals, petroleum products, rubber and paper. The all-industry average encompasses 47 different types of industrial, commercial and housing equipment. Published monthly in each issue of Chemical Engineering until April 2012.
- Intratec Chemical Plant Construction Index, IC: a process plant construction index developed by Intratec, a chemical consulting company. Although cost indexes do not usually forecast future escalation, the IC Index stands out for presenting a smaller delay between release date and index date, besides a 12 months forecast. On the other hand, the index is only available from January 2000 (with a value of 100). A sample is available at Intratec's website.
- Nelson-Farrar Indexes, NF (originally known as the Nelson Refinery Construction Indexes): established in 1946 with a value of 100, the index is more suitable for petroleum or petrochemical business. Published once a month in the Oil and Gas Journal.

Cost Indexes for Different Locations

The standard indexes available are usually restricted to a particular region, as in the case of NF Index, which is accurate only to the region of the US Gulf coast (USGC). Likewise most of the data available for plant and equipment costs are mostly either in

- US Gulf Coast Basis (USGC)
- Northwest Europe Basis (NWE)

Hence the Index values cannot be used to get accurate results in other regions of the world. This variation might be due various factors such as

- Cost and availability of materials
- Cost and availability of labor
- Cost of transportation of equipment and labor
- Import duties and local taxes
- Currency exchange rates

Of all these factors currency exchange rates is said to have the greatest impact in the variation of results.

Cost indexes for international locations is published in *"Process Engineering"* journal as well as in the *"International Journal of Production Economics"* (formerly *"Engineering Costs and Production Economics"*). As of 2012 "*Process Engineering*" has not been publishing new indexes viewable by full access from EbscoHost. They mentioned in May/June 2012 that their analyst had fallen sick.

Location Factor

In order to be able to use the available standard indexes to locations where index data is not available we have to incorporate a new term called the Location Factor (LF) to the standard index value. It is a dimensionless value for a particular location relative to either of the above-mentioned basis.

Cost in A = Cost in USGC x LF(A)

where A is the location for which cost is being evaluated and LF(A) is the location factor for the location A relative to USGC

Location factors are greatly influenced by currency exchange rates due to their significant effect on Index value and hence vary drastically with time. Over the past couple of decades the location factors for various locations are trending close to the value 1. However location factor for a particular region within a country can be easily determined by adding 10% to the index of the reference location for every 1000 miles. The reference location is usually a major industrial location closest to the location where the index is being determined. Location factors for various locations have been published and updated in various journals as in Aspen Richardson's *"International Construction Cost Factor Location Manual (2003)"*.

FINE CHEMICAL

Fine chemicals are complex, single, pure chemical substances, produced in limited quantities in multipurpose plants by multistep batch chemical or biotechnological processes. They are described by exacting specifications, used for further processing within the chemical industry and sold for more than $10/. The class of fine chemicals is subdivided either on the basis of the added value (building blocks, advanced intermediates or active ingredients), or the type of business transaction, namely standard or exclusive products.

Fine chemicals are produced in limited volumes (< 1000 tons/year) and at relatively high prices (> $10/kg) according to exacting specifications, mainly by traditional organic synthesis in multipurpose chemical plants. Biotechnical processes are gaining ground. The global production value is about $85 billion. Fine chemicals are used as starting materials for specialty chemicals, particularly pharmaceuticals, biopharmaceuticals and agrochemicals. Custom manufacturing for the life science industry plays a big role; however, a significant portion of the fine chemicals total production volume is manufactured in house by large users. The industry is fragmented and extends from small, privately owned companies to divisions of big, diversified chemical enterprises. The term "fine chemicals" is used in distinction to "heavy chemicals", which are produced and handled in large lots and are often in a crude state.

commodities	fine chemicals	specialities
single pure chemical substances ...	single pure chemical substances	mixtures
produced in dedicated plants	produced in multi-purpose plants	formulated
high volume / low price	low vol. (< 1000 mtpa) high price (> $ 10/kg)	undifferentiated
many applications	few applications	undifferentiated
sold on specifications	sold on specifications „what they are“	sold on performance „what they can do“

Definition of Fine Chemicals (as opposed to Commodities and Specialties).

Since their inception in the late 1970s, fine chemicals have become an important part of the chemical industry. The total production value of $85 billion is split about 60 / 40 among in-house production by the main consumers, the life science industry, on the one hand, and the fine chemicals industry on the other hand. The latter pursues both a "supply push" strategy, whereby standard products are developed in-house and offered ubiquitously, and a "demand pull" strategy, whereby products or services determined by the customer are provided exclusively on a "one customer / one supplier" basis. The products are mainly used as building blocks for proprietary products. The hardware of the top tier fine chemical companies has become almost identical. The design, lay-out and equipment of the plants and laboratories has become practically the same all over the world. Most chemical reactions performed go back to the days of the dyestuff industry. Numerous regulations determine the way labs and plants have to be operated, thereby contributing to the uniformity.

History

The roots of both the term "fine chemicals" and the emergence of the fine chemical industry as a distinct entity date back to the late 1970s, when the overwhelming success of the histamine H_2 receptor antagonists Tagamet (cimetidine) and Zantac (ranitidine hydrochloride) created a strong demand for advanced organic chemicals used in their manufacturing processes. As the in-house production capacities of the originators, the pharmaceutical companies Smith, Kline & French and Glaxo, could not keep pace with the rapidly increasing requirements, both companies (now merged as GlaxoSmithKline) outsourced part of the manufacturing to chemical companies experienced in producing relatively sophisticated organic molecules. Lonza, Switzerland, which already had supplied an early intermediate, methyl acetoacetate, during drug development, soon became the main supplier of more and more advanced precursors. The signature of a first, simple supply contract is generally acknowledged as the historical document marking the beginning of the fine chemical industry.

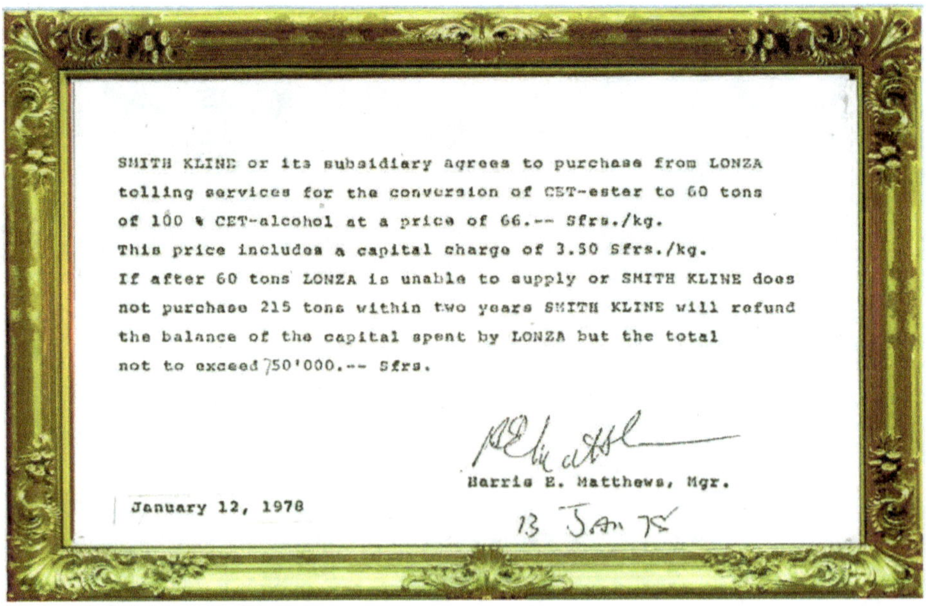

SMITH KLINE or its subsidiary agrees to purchase from LONZA tolling services for the conversion of CET-ester to 60 tons of 100 % CET-alcohol at a price of 66.-- Sfrs./kg.
This price includes a capital charge of 3.50 Sfrs./kg.
If after 60 tons LONZA is unable to supply or SMITH KLINE does not purchase 215 tons within two years SMITH KLINE will refund the balance of the capital spent by LONZA but the total not to exceed 750'000.-- Sfrs.

Harris E. Matthews, Mgr.
13 Jan 78

January 12, 1978

The beginning: supply contract between Smith Kline French and Lonza for cimetidine precursors

In the subsequent years, the business developed favorably and Lonza was the first fine chemical company entering in a strategic partnership with SKF. In a similar way, Fine Organics, UK became the supplier of the thioethyl-N'-methyl-2-nitro-1,1-ethenediamine moiety of ranitidine, the second H2 receptor antagonist, marketed as Zantac by Glaxo. Other pharmaceutical and agrochemical companies gradually followed suit and also started outsourcing the procurement of fine chemicals. An example in case is F.I.S., Italy, which partnered with Roche, Switzerland for custom manufacturing precursors of the benzodiazepine class of tranquilizers, such as Librium (chlordiazepoxide HCl) and Valium (diazepam).

The growing complexity and potency of new pharmaceuticals and agrochemicals requiring production in multipurpose, instead of dedicated plants and, more recently, the advent of biopharmaceuticals had a major impact on the demand for fine chemicals and the evolution of the fine chemical industry as a distinct entity. For many years, however, the life science industry continued considering captive production of the active ingredients of their drugs and agrochemicals as a core competency. Outsourcing was recurred to only in exceptional cases, such as capacity shortfalls, processes requiring hazardous chemistry or new products, where uncertainties subsisted about the chance of a successful launch.

Products

In terms of molecular structure, one distinguishes first between low-molecular-weight (LMW) and high-molecular-weight (HMW) products. The generally accepted threshold between LMW and HMW is a molecular weight of about 700. LMW fine chemicals, also designated as small molecules, are produced by traditional chemical synthesis, by microorganisms (fermentation or biotransformation) or by extraction from plants and animals. In the production of modern life science products, total synthesis from petrochemicals prevails. The HMW products, respectively large molecules, are obtained mainly by biotechnology processes. Within LMWs, the N-heterocyclic compounds are the most important category; within HMWs they are the peptides and proteins.

Small Molecules

As aromatic compounds have been exhausted to a large extent as building blocks for life science products, N-heterocyclic structures prevail nowadays. They are found in many natural products, such as chlorophyll; hemoglobin; and the vitamins biotin, folic acid, niacin (PP), pyridoxine (vitamin B_6), riboflavin (vitamin B_2), and thiamine (vitamin B_1). In synthetic life science products, N-heterocyclic moieties are widely diffuses both pharmaceuticals and agrochemicals. Thus, β-lactams are structural elements of penicillin and cephalosporin antibiotics, imidazoles are found both in modern herbicides, e.g. Arsenal (imazapyr) and pharmaceuticals, e.g. the antiulcerants Tagamet (cimetidine. see above) and Nexium (omeprazole), the antimycotics Daktarin (miconazole), Fungarest (ketoconazole) and Travogen (isoconazole). Tetrazoles and tetrazolidines are pivotal parts of the "sartan" class of hypertensives, e.g. Candesartan cilexetil (candesartan), Avapro (irbesartan), Cozaar (losartan) and Diovan (valsartan).

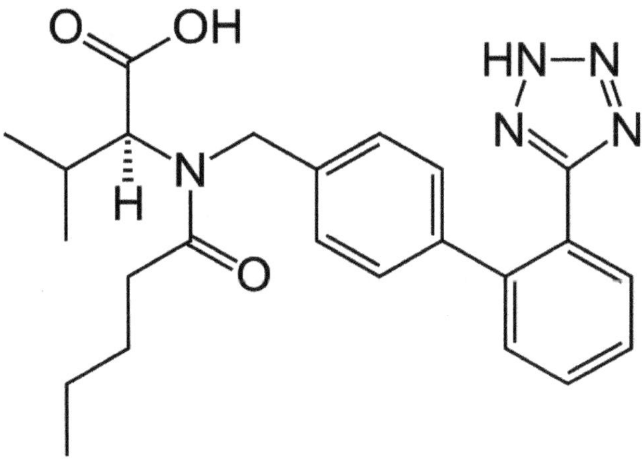

Chemical Structure of Diovan (valsartan)

A vast array of pharmaceuticals and agrochemicals are based on pyrimidines, such as Vitamin B1 (thiamine), the sulfonamide antibiotics, e.g. Madribon (sulfadimethoxime) and –half a century later– the sulfonyl urea herbicides, e.g. Eagle (amidosulfuron) and Londax (bensulfuron-methyl). Benzodiazepine derivatives are the pivotal structural elements of breakthrough CNS drugs, such as Librium (chlordiazepoxide) and Valium (diazepam). Pyridine derivatives are found in both well-known Diquat and Chlorpyrifos herbicides, and in modern nicotinoid insecticides, such as Imidacloprid. Even modern pigments, such as diphenylpyrazolopyrazoles, quinacridones, and engineering plastics, such as polybenzimidazoles, polyimides, and triazine resins, exhibit an N-heterocyclic structure.

Big Molecules

Big molecules, also called *high molecular weight*, HMW molecules, are mostly oligomers or polymers of small molecules or chains of amino acids. Thus, within pharma sciences, peptides, proteins and oligonucleotides constitute the major categories. *Peptides and proteins* are oligomers or polycondensates of amino acids linked together by a carboxamide group. The threshold between the two is as at about 50 amino acids. Because of their unique biological functions, a significant and growing part of new drug discovery and development is focused on this class of biomolecules. Their biological functions are determined by the exact arrangement or sequence of different amino acids in their makeup. For the synthesis of peptides, four categories of fine chemicals, commonly referred to as peptide building blocks (PBBs), are key, namely amino acids (=starting materials), protected amino acids, peptide fragments and peptides themselves. Along the way, the molecular weights increase from about 10^2 up to 10^4 and the unit prices from about \$100 up to \$$10^5$ per kilogram. However, only a small part of the total amino acid production is used for peptide synthesis. In fact, L-glutamic acid, D, L-methionine, L-aspartic acid and L-phenylalanine are used in large quantities as food and feed additives. About 50 peptide drugs are commercialized. The number of amino acids that make up a specific peptide varies widely. At the low end are the dipeptides. The most important drugs with a dipeptide (L-alanyl-L-proline) moiety are the "-pril" cardiovascular drugs, such as Alapril (lisinopril), Captoril (captopril), Novolac (imidapril) and Renitec (enalapril). Also the artificial sweetener Aspartame (N-L-α-Aspartyl-L-phenylalanine 1-methyl ester) is a dipeptide. At the high end there is the anticoagulant hirudin, MW $\approx$ 7000, which is composed of 65 amino acids.

Apart from pharmaceuticals, peptides are also used for diagnostics and vaccines. The total production volume (excl. Aspartame) of chemically synthesized, pure peptides is about 1500 kilograms and sales approach $500 million on the active pharmaceutical (API) level and $10 billion on the finished drug level, respectively. The bulk of the production of peptide drugs, which comprise also the first generation anti-AIDS drugs, the "...navirs", is outsourced to a few specialized contract manufacturers, such as Bachem, Switzerland; Chengu GT Biochem, China; Chinese Peptide Company, China; Lonza, Switzerland, and Polypeptide, Denmark. *Proteins* are "very high-molecular-weight" (MW > 100'000) organic compounds, consisting of amino acid sequences linked by peptide bonds. They are essential to the structure and function of all living cells and viruses and are among the most actively studied molecules in biochemistry. They can be made only by advanced biotechnological processes; primarily mammalian cell cultures. Monoclonal antibodies (mAb) prevail among human-made proteins. About a dozen of them are approved as pharmaceuticals. Important modern products are EPO (Binocrit, NeoRecormon, erythropoietin), Enbrel (etanercerpt), Remicade (infliximab); MabThera/Rituxin (rituximab), and Herceptin (trastuzumab). PEGylation is a big step forward regarding administration of peptide and protein drugs. The method offers the two-fold advantage of substituting injection by oral administration and reducing the dosage, and therefore the cost of the treatment. The pioneer company in this field is Prolong Pharmaceuticals which has developed a PEGylated erythropoietin (PEG-EPO).

Oligonucleotides are a third category of big molecules. They are oligomers of nucleotides, which in turn are composed of a five-carbon sugar (either ribose or desoxyribose), a nitrogenous base (either a pyrimidine or a purine) and 1–3 phosphate groups. The best known representative of a nucleotide is the coenzyme ATP (=Adenosine-triphosphate), MW 507.2. Oligonucleotides are chemically synthesized from protected phosphoramidites of natural or chemically modified nucleosides. The oligonucleotide chain assembly proceeds in the direction from 3'- to 5'-terminus by following a procedure referred to as a "synthetic cycle". Completion of a single synthetic cycle results in the addition of one nucleotide residue to the growing chain. The maximum length of synthetic oligonucleotides hardly exceeds 200 nucleotide components. From its current range of applications in basic research, a.o. in drug target validation, drug discovery and therapeutic development, the potential use of oligonucleotides is foreseen in gene therapy (antisense drugs), disease prevention and agriculture.

Antibody-drug conjugates (ADC) constitute a combination between small and big molecules. The small molecule parts, up to four different APIs, are highly potent cytotoxic drugs. They are linked with a monoclonal antibody, a big molecule which is of little or no therapeutic value in itself, but extremely discriminating for its targets, the cancer cells. The first commercialized ADCs were Isis's Formivirisen and, more recently, Pfizer's (formerly Wyeth) Mylotarg (gemtuzumab ozogamicin). Examples of ADCs in phase III of development are Abbott's / Isis's Alicaforsen and Eli Lilly's Aprinocarsen.

Technologies

Several key technologies are used for the production of fine chemicals, including

- Chemical synthesis, either from petrochemical starting materials or from natural products extracts
- Biotechnology, for small molecules biocatalysis (enzymatic methods), biosynthesis (fermentation), and, for big molecules, cell culture technology.
- Extraction from animals, microorganisms, or plants; isolation and purification, used, for example, for alkaloids, antibacterials (especially penicillins), and steroids
- Hydrolysis of proteins, especially when combined with ion exchange chromatography, used, for instance, for amino acids.

Chemical synthesis and biotechnology are most frequently used; sometimes also in combination.

Traditional Chemical Synthesis

A large toolbox of chemical reactions is available for each step of the synthesis of a fine chemical. The reactions have been developed on laboratory scale by academia over the last two centuries and subsequently adapted to industrial scale, for instance for the manufacture of dyestuffs & pigments. The most comprehensive handbooks describing organic synthetic methods is *Methods of Molecular Transformations*. About 10% of the 26’000 synthetic methods described in are currently used on an industrial scale for fine chemicals production. Amination, condensation, esterification, Friedel–Crafts, Grignard, halogenation (esp. chlorination), and hydrogenation, respectively reduction (both catalytic and chemical) are most frequently mentioned on the websites of individual companies. Optically active cyanohydrins, cyclopolymerization, ionic liquids, nitrones, oligonucletides, peptide (both liquid- and solid-phase), electrochemical reactions (e.g., perfluorination) and steroid synthesis are promoted by only a limited number of companies. With the exception of some stereospecific reactions, particularly biotechnology, mastering these technologies does not represent a distinct competitive advantage. Most reactions can be carried out in standard multipurpose plants. The very versatile organometallic reactions (e.g., conversions with lithium aluminum hydride, boronic acids) may require temperatures as low as -100 °C, which can be achieved only in special cryogenic reaction units, either by using liquefied nitrogen as coolant or by installing a low-temperature unit. Other reaction-specific equipment, such as filters for the separation of catalysts, ozone or phosgene generators, can be purchased in many different sizes. The installation of special equipment generally is not a critical path on the overall project for developing an industrial-scale process of a new molecule.

Since the mid-1990s the commercial importance of *single-enantiomer* fine chemicals has increased steadily. They constitute about half of both existing and developmental drug APIs. In this context, the ability to synthesize chiral molecules has become an important competency. Two types of processes are used, namely the physical separation of the enantiomers and the stereo specific synthesis, using chiral catalysts. Among the latter, enzymes and synthetic BINAP (2,2′–Bis(diphenylphosphino)–1,1′–binaphthyl) types are used most frequently. Large volume (> 103 mtpa) processes using chiral catalysts include the manufacture of the perfume ingredient l-Menthol and Syngenta's Dual (metolachlor) as well as BASF's Outlook (dimethenamid-P) herbicides. Examples of originator drugs, which apply asymmetric technology, are AstraZeneca's Nexium (esomeprazole) and Merck & Co's Januvia (sitagliptin). The physical separation of chiral mixtures and purification of the desired enantiomer can be achieved either by classical fractional crystallization (having a "low-tech" image but still widely used), carried-out in standard multipurpose equipment or by various types of chromatographical separation, such as standard column, simulated moving-bed (SMB) or supercritical fluid (SCF) techniques. For *peptides* three main types of methods are used, namely chemical synthesis, extraction from natural substances, and biosynthesis. Chemical synthesis is used for smaller peptides made of up to 30–40 amino acids. One distinguishes between "liquid phase" and "solid phase" synthesis. In the latter, reagents are incorporated in a resin that is contained in a reactor or column. The synthesis sequence starts by attaching the first amino acid to the reactive group of the resin and then adding the remaining amino acids one after the other. In order to ascertain a full selectivity, the amino groups have to be protected in advance. Most developmental peptides are synthesized by this method, which lends itself to automation. As the intermediate products resulting from individual synthetic steps cannot be purified, a virtually 100% selectivity is essential for the synthesis of larger-peptide molecules. Even at a selectivity of 99% per reaction step, the purity will drop to less than 75% for a dekapeptide (30 steps). Therefore, for industrial quantities of peptides not more than 10–15 amino acid peptides can be made using the solid-phase method. For laboratory quantities, up to 40 are possible. In order to prepare larger peptides, individual fragments are first produced, purified, and then combined to the final molecule by liquid phase synthesis. Thus, for the production of Roche's anti-AIDS drug Fuzeon (enfuvirtide), three fragments of 10–12 amino acids are first made by solid-phase synthesis and then linked together by liquid-phase synthesis. The preparation of the whole 35 amino acid peptide requires more than 130 individual steps.

Microreactor *Technology* (MRT), making part of "process intensification", is a relatively new tool that is being developed at several universities, as well as leading fine chemical companies, such as Bayer Technology Services, Germany; Clariant, Switzerland; Evonik-Degussa, Germany; DSM, The Netherlands; Lonza, Switzerland; PCAS, France, and Sigma-Aldrich, US. The latter company produces about 50 fine chemicals up to multi-kilogram quantities in microreactors. From a technological point of view, MRT, a.k.a. continuous flow reactors, represents the first breakthrough development in reactor design since the introduction of the stirred-tank reactor, which was used by Perkin & Sons, when they set up a factory on the banks of what was then the Grand Junction Canal in London in 1857 to produce mauveïne, the first-ever synthetic purple dye. For a comprehensive coverage of the subject. Examples for reactions that have worked in microreactors include aromatics oxidations, diazomethane conversions, Grignards, halogenations, hydrogenations, nitrations and Suzuki couplings. According to experts in the field, 70% of all chemical reactions could be done in microreactors, however only 10-15% are economically justified.

With the exception of some stereospecific reactions, particularly biotechnology, mastering these technologies does not represent a distinct competitive advantage. Most reactions can be carried out in standard multipurpose plants. Reaction-specific equipment, such as ozone or phosgene generators, is readily available. The installation generally is not a critical path on the overall project for developing an industrial-scale process of a new molecule.

Whereas the overall demand for outsourced pharmaceutical fine chemicals is expected to increase moderately, the estimated annual growth rates for the above-mentioned niche technologies are much higher. Microreactors and the SMB separation technology are expected to grow at a rate of even 50–100% per year. However, the total size of the accessible market typically does not exceed a few hundred tons per year at best.

Biotechnology

Industrial biotechnology, also called *"white biotechnology"* is increasingly impacting the chemical industry, enabling both the conversion of renewable resources, such as sugar or vegetable oils, and the more efficient transformation of conventional raw materials into a wide range of commodities (e.g., cellulose, ethanol and succinic acid), fine chemicals (e.g. 6-aminopenicillanic acid), and specialties (e.g., food and feed additives). As opposed to green and red biotechnology, which relate to agriculture and medicine, respectively, white biotechnology enables the production of existing products in a more economic and sustainable fashion on the one hand, and provides access to new products, especially biopharmaceuticals, on the other hand. It is expected that revenues from white biotechnology will account for 10%, or $250 billion, of the global chemical market of $2,500 billion by 2013. In ten to 15 years it is expected that most amino acids and vitamins and many specialty chemicals will be produced by means of biotechnology Three very different process technologies -biocatalysis, biosynthesis (microbial fermentation), and cell cultures- are used.

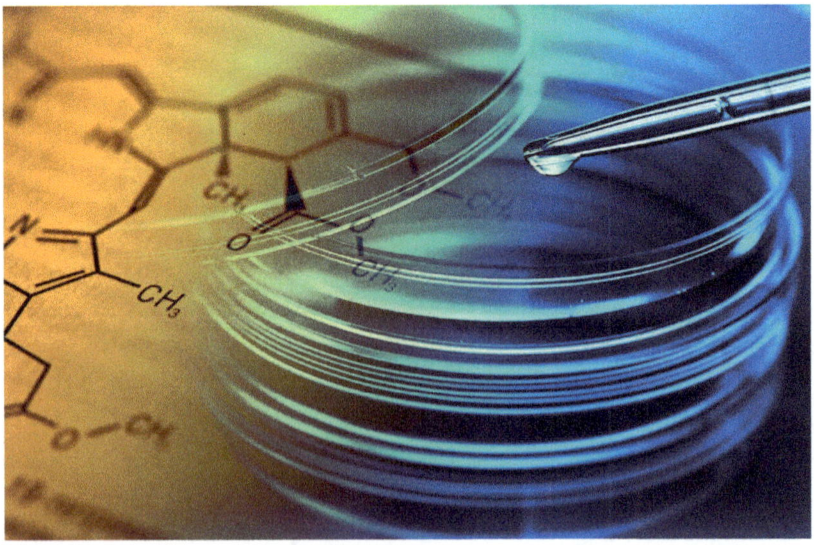

Biocatalysis, a.k.a. biotransformation and bioconversion, makes use of natural or modified isolated enzymes, enzyme extracts, or whole-cell systems for enhancing the production of small molecules. It has much to offer compared to traditional organic synthesis. The syntheses are shorter, less energy intensive and generate less waste and hence, are both environmentally and economically more attractive. About 2/3 of chiral products produced on large industrial scale are already made using biocatalysis. In the manufacture of fine chemicals, enzymes represent the single most important technology for radical cost reductions. This is particularly the case in the synthesis of molecules with chiral centres. Here, it is possible to substitute the formation of a salt with a chiral compound, e.g., (+)-α-phenylethylamine, crystallization, salt breaking and recycling of the chiral auxiliary, resulting in a theoretical yield of not more than 50%, with a one step, high yield reaction under mild conditions and resulting in a product with a very high enantiomeric excess (ee). An example is AstraZeneca's blockbuster drug Crestor (rosuvastatin).

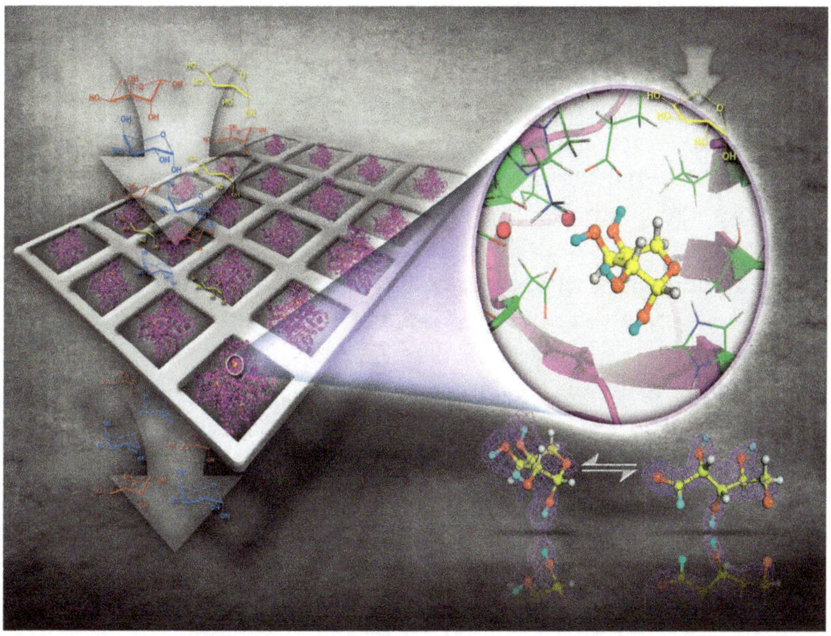

Chemical / Enzymatic Synthesis of Crestor (rosuvastatin)

Further examples of modern drugs, where enzymes are used in the synthesis, are Pfizer's Lipitor (atorvastatin), where the pivotal intermediate R-3-Hydroxy-4-cyanobutyrate is now made with a nitrilase, and Merck & Co.'s Singulair (montelukast), where the reduction of a ketone to S-alcohol, which had required stoichiometric amounts of expensive and moisture sensitive "(-)-DIP chloride" is now replaced by a ketoreductase enzyme catalyst step. Similar rewarding switches from chemical steps to enzymatic ones have also been achieved in steroid synthesis. Thus, it has been possible to reduce the number of steps required for the synthesis of Dexamethasone from bile from 28 to 15. Enzymes differ from chemical catalysts particularly with regard to stereoselectivity, regioselectivity, and chemoselectivity. They can also be modified ("reshuffled") for specific reactions, for use in chemical synthesis. "Immobilized enzymes" are those fixed on solid supports.

They can be recovered by filtration after completion of the reaction. Conventional plant equipment can be used with no, or only modest, adaptations. The *International Union of Biochemistry and Molecular Biology* (IUBMB) has developed a classification for enzymes. The main categories are Oxidoreductases, Transferases, Hydrolases, Lipases (subcategory), Lyases, Isomerases and Ligases, Companies specializing in making enzymes are Novozymes, Danisco (Genencor). Codexis is the leader in modifying enzymes to specific chemical reactions. The highest-volume chemicals made by biocatalysis are bio-ethanol (70 million metric tons), high-fructose corn syrup (2 million metric tons); acrylamide, 6-aminopenicillanic acid (APA), L-lysine and other amino acids, citric acid and niacinamide (all more than 10'000 metric tons).

Biosynthesis i.e. the conversion of organic materials into fine chemicals by microorganisms, is used for the production of both small molecules (using enzymes in whole cell systems) and less complex, non-glycosylated big molecules, including peptides and simpler proteins. The technology has been used for 10,000 years to produce food products, like alcoholic beverages, cheese, yogurt, and vinegar. In contrast to biocatalysis, a biosynthetic process does not depend on chemicals as starting materials, but only on cheap natural feedstock, such as glucose, to serve as nutrient for the cells. The enzyme systems triggered in the particular microorganism strain lead to the excretion of the desired product into the medium, or, in the case of HMW peptides and proteins, to the accumulation within so-called inclusion bodies in the cells. The key elements of fermentation development are strain selection and optimization, as well as media and process development. Dedicated plants are used for large-scale industrial production. As the volume productivity is low, the bioreactors, called fermenters, are large, with volumes that can exceed 250 m3. Product isolation was previously based on large-volume extraction of the medium containing the product. Modern isolation and membrane technologies, like reverse osmosis, ultra- and nano-filtration, or affinity chromatography can help to remove salts and by-products, and to concentrate the solution efficiently and in an environmentally friendly manner under mild conditions. The final purification is often achieved by conventional chemical crystallization processes. In contrast to the isolation of small molecules, the isolation and purification of microbial proteins is tedious and often involves a number of expensive large-scale chromatographic operations. Examples of large-volume LMW products made by modern industrial microbial biosynthetic processes are monosodium glutamate (MSG), vitamin B2 (riboflavin), and vitamin C (ascorbic acid). In vitamin B2, riboflavin, the original six- to eight-step synthetic process starting from barbituric acid has been substituted completely by a microbial one-step process, allowing a 95% waste reduction and an approximately 50% manufacturing cost reduction. In ascorbic acid, the five-step process (yield $\approx$ 85%) starting from D-glucose, originally invented by Tadeus Reichstein in 1933, is being gradually substituted by a more straightforward fermentative process with 2-ketogluconic acid as pivotal intermediate. After the discovery of penicillin in 1928 by Sir Alexander Fleming from colonies of the bacterium Staphylococcus aureus, it took more than a decade before a powdery form of the medicine was developed. Since then, many more antibiotics and other secondary metabolites have been isolated and manufactured by microbial fermentation on a large scale. Some important antibiotics besides penicillin are cephalosporins, azythromycin, bacitracin, gentamycin, rifamycin, streptomycin, tetracycline, and vancomycin.

Cell Cultures Animal or plant cells, removed from tissues, will continue to grow if cultivated under the appropriate nutrients and conditions. When carried out outside the natural habitat, the process is called cell culture. *Mammalian cell culture* fermentation, also known as recombinant DNA technology, is used mainly for the production of complex big molecule therapeutic proteins, a.k.a. biopharmaceuticals. The first products made were interferon (discovered in 1957), insulin, and somatropin. Commonly used cell lines are Chinese hamster ovary (CHO) cells or plant cell cultures. The production volumes are very small. They exceed 100 kg per year for only three products: Rituxan (Roche-Genentech), Enbrel (Amgen and Merck & Co. [formerly Wyeth]), and Remicade (Johnson & Johnson). Fine chemical production by mammalian cell culture is a much more demanding operation than conventional biocatalysis and –synthesis. The bioreactor batch requires more stringent controls of operating parameters, since mammalian cells are heat and share sensitive; in addition the growth rate of mammalian cells is very slow, lasting from days to several months. While there are substantial differences between microbial and mammalian technologies (e.g., the volume / value relationships are 10 $/kg and 100 tons for microbial, 1'000'000 $/kg and 10 kilograms for mammalian technology; the cycle times are 2–4 and 10–20 days, respectively), they are even more pronounced between mammalian and synthetic chemical technology.

	Mammalian cell technology	**Chemical technology**
Worldwide reactor volume	≈ 3'000 m^3 (fermenters)	≈ 80'000 m^3
Investment per m^3 reactor volume	≈ $5 million	≈ $500'000
Production per m^3 reactor volume and year	several 10 kg	several 1'000 kg
Sales per m^3 reactor volume and year	≈ $5 – 10 million	≈ $250'000 - 500'000
Value of 1 batch	≈ $5 million (20'000 liter fermenter)	≈ $500'000
Product concentration in reaction mixture	≈ 2 - 6 (-10) g / Liter	≈ 100 g / Liter (10%)
Typical reaction time	≈ 20 days	≈ 6 hours
Process development time	≈ 3 years (one step)	2 – 3 months per step
Capacity expansion projects	many, doubling of actual capacity	few, mainly in Far East
Governing rules	cGMP, BLA [Biological License Application (product specific)]	cGMP, ISO 14000
Scale-up factor (1st lab process to industrial scale)	≈ 10^9 (μg → 1 ton)	≈ 10^6 (10 g → 10 tons)
Plant construction time	4 – 6 years	2 – 3 years

share of outsourcing	early stage	55%	25% of chemical production
	commercial	20%	45% of chemical production

The mammalian cell production process, as used for most biopharmaceuticals, is divided into the four main steps: (1) Cultivation, i.e. reproduction of the cells; (2) Fermentation, i.e. the actual production of the protein, typically in 10'000 Liter, or multiples, bioreactors; (3) Purification, i.e. separation of the cells from the culture medium and purification, mostly by chromatography, (4) Formulation, i.e. conversion of the sensitive proteins to a stable form. All steps are fully automated. The low productivity of the animal culture makes the technology expensive and vulnerable to contamination. Actually, as a small number of bacteria would soon outgrow a larger population of animal cells. Its main disadvantages are low volume productivity and the animal provenance. It is conceivable that other technologies, particularly plant cell production, will gain importance in future. Given the fundamental differences between the two process technologies, plants for mammalian cell culture technologies have to be built ex novo.

The pro's and con's of an involvement of a fine chemical company in cell culture technology are listed below:

Pros:

- Strong growth of demand: Today, biopharmaceuticals account for about \$55–\$80 billion, or 15% of the total pharmaceutical market. They are growing by 15% per year, i.e. three times faster than LMW drugs and are expected to pass the \$150 billion per year threshold by 2015. Whereas just one out of the world's top ten drugs was a biopharmaceutical in 2001, the number went up to five in 2010 and is expected to increase further to eight by 2016 .

Proprietary name		Generic Name	Company
Small Molecular Weight (conventional chemical)			
1	Crestor	rosuvastatin	AstraZeneca
2	Advair/Seretide	Salmeterol / fluticasone	GlaxoSmithKline
High Molecular Weight (biopharmaceuticals)			
1	Humira	adalimumab	AbbVie (Before: Abbott)
2	Enbrel	etanecept	Amgen
3	Prolia	denosumab	Amgen
4	Rituxan	rituximab	Roche/Biogen Idec
5	Avastatin	bevacizumab	Roche
6	Herceptin	trastuzumab	Roche
7	Remicade	infliximab	J&J/Merck & Co.
8	Lantus	insulin glargine	Sanofi-Aventis

The likelihood of developing a new biopharmaceutical successfully is significantly greater than in traditional drug development. 25% of biopharmaceuticals that enter Phase I of the regulatory process eventually are granted approval. The corresponding figure for conventional drugs is less than 6%.

- The traditionally large share of outsourcing.
- Small number of custom manufacturers with industrial-scale manufacturing capabilities in this demanding technology. In the Western hemisphere, primarily Boehringer-Ingelheim of Germany and Lonza of Switzerland; in the Eastern hemisphere Nicholas Piramal of India (through the acquisition of a former Avecia operation)and the joint ventures between AutekBio and Beijing E-Town Harvest International in China and between Biocon in India and Celltrion in South Korea.
- Same customer category: life science, especially the pharmaceutical industry.
- Similar business types: custom manufacturing of proprietary drugs; opportunities for generic versions, called biosimilars.
- Similar regulatory environment: FDA regulations, especially GMP.
- Existing infrastructure (utilities, etc.) can be used.

Cons:

- High entry barriers because of demanding technology. The construction of a large-scale plant for the production of biopharmaceuticals by cell culture fermentation costs around $500 billion and takes four to six years.
- As the specifications of the plant and process types for biopharmaceuticals differ substantially from traditional chemical synthesis, they cannot be produced in conventional multipurpose fine chemical plants.
- High financial exposure: (1) high capital intensity ('massive investments are needed at a time when chances of success are still very low' and (2) risk of batch failures (contamination).
- Unlike the biopharmaceutical start-ups, the emerging big biopharmaceutical companies are adopting the same opportunistic outsourcing policy as larger pharmaceutical companies. Thus, Amgen, Biogen Idec, Eli Lilly, Johnson & Johnson (J&J), Medimmune, Novartis, Roche/Genentech and Pfizer are investing heavily in in-house manufacturing capacity. With three plants in the US, two in Japan and one each in Germany and Switzerland, Roche has the largest production capacity.
- New developments in expression systems for mammalian and plant cell technology could reduce capacity requirements substantially. Actually, the titer in large-scale mammalian production, actually 2–3 grams/liter. is expected to double to 5–7 by 2015 and once more to 10 by 2020. Furthermore, the widespread application of 'single-use disposable bioprocessing technology', considered by experts as 'the hottest buzz in town'. It advantageously substitutes for stainless steel production trains, at least for short production campaigns.

- New transgenic production systems are emerging. They (e.g. transgenic moss, lemna, fungal or yeast expression systems, transgenic animals and plants, such as tobacco plants possess the potential to become economically and industrially successful.
- Legislation and regulation of biotechnology is not well defined yet and leads to differences in interpretation and other uncertainties. In the US, legislation is not yet in place for biosimilars, the generic counterpart of generics in small molecule pharmaceuticals.

The inherent risks of the mammalian cell technology led several companies to opt out of mammalian cell technology or to substantially reduce their stake. Examples are Cambrex and Dowpharma in the US, Avecia, DSM and Siegfried in Europe and WuXi App Tech in China. In conclusion, biocatalysis should be, or become, part of the technology toolbox of any fine chemical company. Mammalian cell culture fermentation, on the other hand, should be considered only by large fine chemical companies with a full war chest and a long-term strategic orientation.

The Industry

Within the chemical universe, the fine chemical industry is positioned between the commodity, their suppliers, and specialty chemical industries, their customers. Depending on the services offered, there are two types of fine chemical companies. The *Fine Chemical Companies* are active in industrial scale production, both of standard and exclusive products. If the latter prevails, they are referred to as Fine Chemical / *Custom Manufacturing Organizations (CMOs)*. The main assets of the Contract Research Organizations (CROs) are their research laboratories. CRAMS; Contract Research and Manufacturing Organizations are hybrids.

Fine Chemical / Custom Manufacturing Companies

Fine chemical / Custom Manufacturing companies in the narrower sense are active in process scale up, pilot plant (trial) production, industrial-scale exclusive and non-exclusive manufacture and marketing. Their product portfolios comprise exclusive products, produced by custom manufacturing, as main activity, non-exclusive products, e.g. API-for Generics, and standard products. Characteristics are high asset intensity, batch production in campaigns in multipurpose plants, above-industry-average R&D expenditures and close, multi-level and multi-functional relationships with industrial customers. The industry is very fragmented. 2'000 – 3'000 fine chemical companies exist globally, extending from small, "garage-type" outfits in China making just one product, all the way to the big, diversified enterprises, resp. units. The main reason for the fragmentation is the lack of economy of scale (see below).

The industry is subject to a high degree of regulation even more so than the chemical industry as a whole, particularly if pharmaceutical fine chemical production is involved. The most important regulatory authorities are the *(US) Food and Drug Administration (FDA)* and *(Chinese) State Food & Drug Administration (SFDA)*, respectively. Their main responsibilities comprise formulating comprehensive supervision policies ("Good

Manufacturing Practice") and control the implementation, to be in charge of drug registration, draw up criteria for marketing authorization and formulate national essential medicines lists. The European correspondent is the *European Medicines Agency (EMEA)*, which is manly responsible for the scientific evaluation of medicines developed by pharmaceutical companies for use in the European Union. The role of REACH (Registration, Evaluation, Authorization and Restriction of Chemicals) is self-explanatory. The *U.S. Pharmacopeia* codifies quality standards for Active Pharmaceutical Ingredients. As these standards are observed worldwide, they contribute also to the emergence of a uniform world-wide set-up of the top tier fine chemical companies. In terms of size, resources, and complexity of the chemical process technologies mastered, the fine chemical companies can be broadly divided into three segments, each of them accounting for approximately the same turnover, namely about $10 billion. The top tier, about twenty, has sales in excess of $250 million per year. Most are not pure players but divisions or b.u.'s of large, multinational companies. Their share varies between one percent or less for BASF and Pfizer, all the way to 100% for Cambrex, USA; Divi's Laboratories, India, and F.I.S. Italy. All have extensive resources in terms of chemists and other specialists, plants, process knowledge, backwards integration, international presence, etc.

	Company	**Location**	**Sales 2009 ($ million)**	**F.C. unit**	**Sales 2009 ($ million)**	**Remarks**
1	Lonza	Switz.	2'600	Custom. Manuf.	1'370	HMW/LMW~55/45
2	Boehringer-Ingelheim	Germany	18'300	Fine Chem.[1]	950	HMW/LMW=84/16
3	DSM	The Netherlands	11'300	Fine Chem.[1]	850[aE]	
4	Sumitomo Chemicals	Japan	17'420	Fine Chem.[1]	730	incl. some polymer additives
5	Merck KGaA	Germany	11'200	Life Science Solutions	580	#1 in liquid crystals
6	Sigma-Aldrich	USA	2'148	SAFC	570[E]	
7	BASF	Germany	73'000	Fine Chem.[1]	550[2E]	incl. some excipients
8	CSPC Shijiazhuang Pharmaceutical Group	China	1'500	Fine Chem.[1]	550[E]	API-for-Generics, e.g. HIV / AIDS, sartans
9	Lanxess	Germany	7'280	Saltigo	550[E]	a.o. agrochemicals
10	Albemarle	USA	2'005	Fine Chem.[1]	500[2]	a.o. ibuprofen
Total Top Ten					**~7'200**	

[1] as per author's definition

[2] part of the sales do not derive from fine chemicals, e.g., generics, catalysts, excipients

[E] Author's estimate (non figures published by the company) HMW, high molecular weight, LMW, low molecular weight fine chemicals

11.-20.: Jubilant Organosys. India,800[E]/470; Dr. Reddy's, India, 1'370/370; Evonik-Degussa, Germany, 18'900/350[E]; Johnson Matthey, UK 12'500/350; Aurobinda, India 665/340; NCPC, North China Pharmaceutical, China, 718/300[E]; Divi's Laboratories, India, 250/250; Pfizer, US, 50'000/250[E]; Cambrex, US, 235/235; F.I.S., Italy, 230/230

∑11-20 ~ 2'900 million; ∑∑1-20 ~ $10'000 million

note: The first number refers to the total sales, the second one to the fine chemical sales. Both are in $ million

The combined revenues of the top 20 fine chemical companies amounted to $10 billion in 2009, representing about 30% of the figure for the whole industry. The leading companies are typically divisions of large, diversified chemical companies. In terms of geography, 9 of the top 20 are located in Europe, which is recognized as the cradle of the fine chemical industry. This is e.g. the case for the world's #1 company, Lonza, headquartered in Basel. Switzerland. Custom manufacturing prevails in northern Europe; the manufacture of active substances for generics, in southern Europe. The second largest geographic area is Asia, housing 7 of the top 20. With 4 large companies, the US rank last.

Whereas the European and U.S. pharma industry constitutes the main customer base for most fine chemical companies, some have a significant share of products and services for the agrochemical industry. Examples are Archimica, CABB, Saltigo (all Germany), DSM (The Netherlands) and Hikal, India. Several large pharmaceutical companies market fine chemicals as subsidiary activity to their production for captive use, e.g. Abbott, USA; Bayer Schering Pharma, Boehringer-Ingelheim, Germany; Daiichi-Sankyo (after the takeover of Ranbaxy), Japan; Johnson & Johnson, USA; Merck KGaA, Germany; Pfizer (formerly Upjohn), US. Large fine chemical companies, in contrast to mid-sized and small ones, are characterized by

- *A Lack of Economy in Size*. As most fine chemicals are produced in quantities of not more than a few 10 tons per year in multipurpose plants, there is little or no economy of size. The reactor trains of these plants are similar throughout the industry. Regardless of the size of the companies, their main constituents, the reaction vessels, have a median size of the 4–6 m^3. Various products are made throughout a year in campaigns. Therefore, the unit cost per m^3 per hour does practically not vary with the size of the company.
- *A Dichotomy between Ownership and Management*. The company's shares are listed on stock exchanges, and their performance is scrutinized by the financial community. Postponement of a single important shipment can affect a quarterly result. In the small and mid-sized companies the owners typically are the major shareholders, often members of the same family. Their shares are not traded publicly and fluctuations in their financial performance are more easily coped with.
- *Complicated Business Processes*. Flexibility and Responsiveness are in jeopardy. Customer complaints, for instance, are difficult to resolve in a straightforward manner.
- *A Heterogeneous portfolio of small companies, accumulated over time through M&A activities*. The key functions, such as production, R&D, and M&S, are located on different sites, often in different countries.
- *A Cohabitation with Other Units*.

A comprehensive list of about 1400 fine chemical companies (including traders) can be found in the "event catalogue" of the CPhI exhibition.

The *second tier* consists of several dozens of *midsized* companies with sales in the range of $100–$250 million per year. Their portfolios comprise both custom manufacturing and API-for-generics. They include both independents and subsidiaries of major companies. A number of these companies are privately owned and have grown mainly by reinvesting the profits. Examples are Bachem, Switzerland; Dishman, India; F.I.S. and Poli Industria Chimica, Italy; Hikal, India, and Hovione, Portugal. Customers prefer to do business with mid-sized companies, because communications are easier —they typically deal directly with the decision maker— and they can better leverage their purchasing power. The *third tier* includes thousands of *small independents* with sales below $100 million per year. Most of them are located in Asia. They often specialize in niche technologies. The minimum economical size of a fine chemical company depends on the availability of infrastructure. If a company is located in an industrial park, where analytical services; utilities, safety, health, and environmental (SHE) services, and warehousing are readily available, there is practically no lower limit. New fine chemical plants have come on-stream mostly in Far East countries over the past few years. Their annual turnover rate rarely exceeds $25 million. All big and medium-size fine chemical companies have cGMP-compliant plants that are suitable for the production of pharmaceutical fine chemicals. With the exception of biopharmaceuticals, which are manufactured by only a few selected fine chemical companies, the technology toolboxes of all these companies are similar. This means that they can carry out practically all types of chemical reactions. They differentiate on the basis of the breadth and quality of the service offering.

Contract Research Organizations

Contract research organizations (CROs) provide services to the life science industries along product development. There are more than 2000 CROs operating worldwide, representing revenues of more than $20 billion. One distinguishes between "Product" and "Patient" CROs. Whereas the production sites of CMOs are multipurpose plants, allowing for the production of tens to hundreds of tons of fine chemicals, the work places of patient CROs are the test persons (volunteers) for the clinical trials and those of the product CROs are the laboratory benches. Major customers for CRO services are the large global pharmaceutical companies. Half a dozen companies (Pfizer, GlaxoSmithKline, Sanofi-Aventis, AstraZeneca, Johnson & Johnson, and Merck & Co.) alone absorb about one third of all CRO spending. As for CMOs also for CROs, biotech start-up companies with their dichotomy between ambitious drug development programs and limited resources are the second most promising prospects. Product CROs (chemical CROs) are providing primarily sample preparation, process research and development services. An overlap between the latter and CMOs exists with regard to pilot plants (100 kg quantities), which are part of the arsenal of both types of enterprise. There are more 100 product CROs. Most of them are privately held and have revenues of $10–$20 million per year or less, adding up to a total business in the range of $1.5-$2 billion. Examples of are:

- In *North America*: Alphora; Delmar; NAEJA, all Canada. AMRI; Aptuit; Cambridge Major; ChemBridge; Innocentive; Irix Pharmaceuticals, PharmEco, all USA.
- In *Europe*; Carbogen-Amcis, Switzerland; Chemcomm, Germany; ChemDiv, Russia; Clauson-Kaas, Denmark; Enamine Ltd, Ukraine; Girindus, Germany; Nerviano Medical Sciences, Italy; Recipharm, Sweden; Serichim, Italy; Solvias, Switzerland.

- In *Asia*: BioDuro, Medicilon, Pharmaron; WuXi AppTec, all China; Acoris; Aptuit Laurus; Biocon / Syngene; Chembiotek; Chempartner; ProCitius, all India; NARD Institute, Riken, both Japan.

The business of CROs is usually done through a "pay for service" arrangement. Contrary to manufacturing companies, invoicing of CROs is not based on unit product price, but on full-time equivalents (FTEs), that is, the cost of a scientist working one year on a given customer assignment. Companies offering both contract research and manufacturing services (CRAMS) combine the activities of CROs and CMOs. Their history is either a forward integration of a CRO, which adds industrial scale capabilities or backwards integration of a CMO. As there are only limited synergies (e.g. > 90% of the projects end at the sample preparation stage). It is questionable, though, whether one-stop shops really fulfil a need. Actually, the large fine chemical companies consider the preparation of samples more as marketing tool (and expense ...) rather than a profit contributor.

The offerings of Patient CROs (Clinical CROs) comprise more than 30 tasks addressing the clinical part of pharmaceutical development at the interface between drugs, physicians, hospitals, and patients, such as the clinical development and selection of lead new drug compounds. As clinical trials represent the largest expense in pharmaceutical research, the market for patient CROs is larger than for their product counterparts. Thus, the sales of the top tier firms, Charles River Laboratories, Covance, Parexel, PPD, Quintiles Transnational, all USA, and TCG Lifescience, India; are in the $1–$2 billion range, whereas the largest product CROs have revenues of a few 100 million dollars.

Research and Development

The overall emphasis of fine chemical R&D is more on development than on research. The main tasks are (1) designing, respectively duplicating and adapting in case of custom manufacture, and developing laboratory procedures for new products or processes; (2) transferring the processes from the laboratory via pilot plant to the industrial scale (the scale up factor from a 10g sample to a 1ton batch is 100,000); and (3) to optimize existing processes. At all times during this course of action it has to be ensured that the four critical constraints, namely, economics, timing, safety, ecology and sustainability are observed . R&D expenditures in the fine chemical industry are higher than in the commodities industry. They represent around 5–10% versus 2–5% of sales. On the business side, product innovation must proceed at a more rapid pace, because life cycles of fine chemicals are shorter than those of commodities. Therefore, there is an ongoing need for substitution of obsolete products. On the technical side, the higher complexity of the products and the more stringent regulatory requirements absorb more resources. Many economic and technical parameters have been proposed to enable a meaningful assessment of single projects and project portfolios. Examples are attractiveness, strategic fit, innovation, gross/net present value, expected profits, R&D expenditures, development stage, probability of success, technology fit, potential conflicts with other activities of the company and realization time. Most of these parameters cannot be determined quantitatively, at least during the early phases of a project. The best way to take advantage of a project portfolio is to develop and use it in an iterative way. By comparing the entries at regular intervals, for instance, every 3 months, the directions that the projects take can be visualized. If a negative trend persists with one particular project, the project should be put on the watch list.

Objectives

R&D has to manage the following functions in order to deliver the requested services: *Literature and Patent Research*. Provisions have to be made for a periodic examination of all acquired research results to safeguard Intellectual Property Rights (IPR) and to determine whether patent applications are indicated. Patent research is particularly important for evaluation of the feasibility of taking up R&D for new APIs-for-generics. *Process Research* has to design new synthetic routes and sequences. Two approaches are feasible. For simple molecules, the "bottom-up" approach is the method of choice. The researcher converts a commercially available starting material and sequentially adds more reagents until the target molecule is synthesized. For more complex molecules, a "top-down" approach, also known as retro synthesis, or de-construction, is chosen. Key fragments of the target molecule are first identified, then synthesized individually, and finally combined to form the desired molecule through convergent synthesis. *Process Development* focuses on the design of new, efficient, stable, safe, and scalable synthetic routes to a target fine chemical. It represents an essential link between process research and commercial production. The resulting "base process" description provides the necessary data for the determination of preliminary raw material and product specifications, the manufacture of semi commercial quantities in the pilot plant, the assessment of the ecological impact, the regulatory submissions and technology transfer to manufacture at industrial scale, and an estimate of the manufacturing costs in an industrial-scale plant. If the base process is provided by the customer as part of the technology transfer, process, research has to optimize it so that it can be transferred to the bench-scale laboratory or pilot plant. Furthermore, it has to be adapted to the specific characteristics of available production trains. *Bench-scale Laboratory, kg-lab and Pilot Plant Development*. Depending on the volume requirements, three different types of equipment are used for process research, development and optimization, namely bench-scale laboratories for gram to 100 gram, kilo-labs for kg to 10 kg and pilot plants for 100 kg to ton quantities. Particularities of laboratory processes that have to be eliminated include the use of large numbers of unit operations, dilute reaction mixtures, vast quantities of solvents for extraction, evaporation to dryness, drying of solutions with hygroscopic salts. Although modern reaction calorimeters consent to foresee the effects of these different conditions to a certain extent, a direct transfer of a process from the laboratory to the industrial scale is not recommended, because of the inherent safety, environmental, and economic risks. In development, the viability of the process on a semi commercial scale has to be demonstrated. Trial quantities of the new fine chemical have to be manufactured for market development, clinical tests, and other requirements. The necessary data have to be generated to enable the engineering department to plan the modifications of the industrial-scale plant and in order to calculate production costs for the expected large-volume requirements. Both equipment and plant layout of the pilot plant reflect those of an industrial multipurpose plant, except for the size of reaction vessels (bench-scale laboratory ~10–60 liters; pilot plant ~100–2500 liters) and the degree of process automation. Before the process is ready for transfer to the industrial-scale plant, the following activities have to be completed: Adaptation of the laboratory process to the constraints of a pilot plant, hazard and operability (HAZOP) analysis, execution of demonstration batches. The main differences between laboratory synthesis and industrial scale production are shown in Table below.

Task	**Laboratory synthesis**	**Industrial scale process**
Operator	Laboratory chemist	Chemical engineer
Economy	Yield	Throughput (kg/m^3/hour)
Units	G, mL, mol; min. hours	Kg, ton, hours, shift
Equipment	Glass flask	Stainless steel, glass lined
Process control	Manual	Automatic [reaction vessel]
Critical path	Reaction time	Heating / cooling
Liquid handling	Pouring	Pumping
Liquid / solid sep.	Filtration	Centrifugation

In case of cGMP fine chemicals also a process validation is required. It consists of the three elements process design, process qualification and continued process verification. *Process Optimization*. Once a new chemical process has been introduced successfully on an industrial scale, process optimization is called upon to improve the economics. As a rule of thumb it should be attempted to reduce the costs of goods sold (COGS) by 10-20%, every time the yearly production quantity has doubled. The task extends from fine tuning the currently used synthetic method all the way to the search for an entirely different second generation process. Specific provisions are the increase of overall yield, the reduction of the number of steps, raw material cost, solvent, catalyst, enzyme consumption, environmental impact.

Project Management

There are two main sources of new research projects, namely ideas originating from the researchers themselves ("supply push") and those coming from customers ("demand pull"). Ideas for new processes typically originate from researchers, ideas for new products from customers, respectively customer contacts. Particularly in custom manufacturing, "demand pull" prevails industrial reality. The "new product committee" is the body of choice for evaluating new and monitoring ongoing research activities. It has the assignment to evaluate all new product ideas. It decides whether a new product idea should be taken up in research, reassesses a project at regular intervals and, last but not least decides also about the abandonment of a project, once it becomes evident that the objectives cannot be reached. In a typical project the overall responsibility for the economic and technical success lies with the project champion. He is assisted by the project manager, who is responsible for the technical success. In custom manufacturing, a typical project starts with the acceptance of the product idea, which originates mainly from business development, by the new product committee, followed by the preparation of a laboratory process, and ends with the successful completion of demonstration runs on industrial scale and the signature of a multiyear supply contract, respectively. The input from the customer is contained in the "technology package". Its main constituents are (1) reaction scheme, (2) target of project & deliverables (product, quantity, required dates, specifications), (3) list of analytical methods, (4) process development opportunities (stepwise assessment), (5) list of required reports, (6) Safety, Health and Environment (SHE) issues, (7) materials to be supplied by customer and (8) packaging & shipping information The technical part of a project usually determines its duration. Depending on the quality of the information contained in the "technology package" received from the customer and the complexity of the project as such, particularly the number of steps that have to be performed; it can be any time between 12 and 24 months. Depending on the number of researches involved, the total budget easily amounts to several million US dollars.

Markets

Fine chemicals are used as starting materials for specialty chemicals. The latter are obtained either by direct formulation or after chemical/biochemical transformation of intermediates to active substances. Life sciences, primarily pharmaceutical, agrochemical and food and feed industries are the main consumers of fine chemicals.

Market size

Fine chemicals account for about 4% of the universe of chemicals. The latter, valued at $2'500 billion, is dominated mainly by oil-, gas-, and mineral-derived commodities (~40%) on one hand and a large variety of specialty chemicals at the interface between industry and the public on the other hand (~55%). The global production value of fine chemicals is estimated at $85 billion, of which about 2/3, or $55 billion are produced captively and $30 billion represent the global revenues of the fine chemical industry. The corresponding figures for the major user, the pharmaceutical industry, are $32 billion and $23 billion, respectively. For a number of reasons, such as the lack of statistical data and the somewhat equivocal definition it is not possible to exactly determine the size of the fine chemical market.

total A.I.		Size ($ billion)		
		captive	merchant	
Life sciences	Pharmaceuticals	55	32	23
	Agrochemicals	15	11	4
Various specialty chemicals		15	10	5
Total fine-chemical industry		85	53	32

In this table, the approximately $85 billion fine chemical market is subdivided into major applications according to their relevance, namely, fine chemicals for pharmaceuticals, agrochemicals and specialty chemicals outside life sciences. Furthermore, a distinction is made between captive (in-house) production and merchant market. Pharmaceutical fine chemicals (PFCs) account for two-thirds of the total. Out of the PFC value of $55 billion, about $23 billion (~40%) are traded, and $32 billion (~60%) are the production value of the pharma industry's in-house production. Within life science products, fine chemicals for agro, and —at a distance— for veterinary drugs follow in importance. The production value for fine chemicals used for specialty chemicals other than pharmaceuticals and agrochemicals is estimated at $15 billion. As the leading specialty chemical companies, Akzo Nobel, Dow, Du Pont, Evonik, Chemtura and Mitsubishi are backward-integrated, the share of in-house production is estimated at 75%, leaving a merchant market of approximately $5 billion.

Target Markets

Pharmaceuticals

The pharmaceutical industry constitutes the most important customer base for the fine chemical industry. The largest companies are Pfizer, USA; Roche, Switzerland, GlaxoSmithKline, UK; Sanofi Aventis, France, and Novartis, Switzerland. All are active in R&D, manufacturing and marketing. Pharmaceuticals containing more than 2'000 different active ingredients are in commerce today; a sizable number of them are sourced from the fine chemical industry. The industry also has a track record of above-average growth. The fine chemical industry has a keen interest in the top-selling or "blockbuster drugs", i.e. those with worldwide annual sales in excess of $1 billion. Their number has increased steadily, from 27 in 1999 to 51 in 2001, 76 in 2003, and than levelled off.

	Brand	**API**	**Company**	**sales 2010 ($ bio)**
1	Lipitor	atorvastatin	Pfizer	11.8
2	Plavix	clopidogrel	Bristol-Myers Squibb Sanofi-Aventis	9.4
3	*Remicade**	*infliximab*	J&J, Merck, Mitsubishi, Tanabe	8.0
4	Advair/ Seretide	salmeterol + fluticasone	Glaxo SmithKline	8.0
5	*Enbrel**	*etanecerpt*	Amgen, Pfizer, Takeda	7.4
6	*Avastin**	*bevacizumab*	Roche	6.8
7	Abilify	*aripiprazole*	Bristol-Myers Squibb Otsuka	6.8
8	*Mabthera/ Rituxan**	*rituximab*	Roche	6.7
9	*Humira**	*adalimubab*	AbbVie (Before: Abbott)	6.5
10	Diovan & Co-Diovan	valsartan	Novartis	6.1
Total Top 10				77.5

Sales of the top 20 blockbuster drugs are reported in this table . The APIs of 12 of them are "small" (LMW) molecules. Averaging a MW of 477, they have quite complex structures. They typically show three cyclic moieties. 10 of them exhibit at least one N-heterocyclic moiety. Five of the top 10, up from none in 2005, are biopharmaceuticals. The largest-selling non-proprietary drugs are paracetamol, omeprazole, ethinylestradiol, amoxicillin, pyridoxine, and ascorbic acid. The innovator pharma companies require mainly custom manufacturing services for their proprietary drug substances. The demand is driven primarily by the number of new drug launches, the volume requirements and the industry's "make or buy" strategy. A summary of the pro's and con's for outsourcing from the pharma industry's perspective is given in Table below. As extended studies at the Stern Business School of the New York City University have shown, financial considerations clearly favor the "buy" option.

Pro's	Con's
concentrate on core activities (innovation & marketing)	loss of tax benefits resulting from production of APIs in tax havens
deploy your financial resources for more profitable investments	dissemination of Intellectual Property
benefit from F.C. industry's know how and expertise	loss of know how
eliminate long lead times to build and validate a manufacturing facility	job losses
free capacity for new product introductions	under-utilization of in-house production capacity
avoid risks of using hazardous chemistry	

Teva and Sandoz are by far the largest *generics companies*. They differ from their competitors not only in sales revenues but also because they are strongly backwards integrated and have proprietary drugs in their portfolios. They also vie for the promising biosimilars market.

Several thousand *small* or *virtual pharma* companies focus on R&D. albeit on just a few lead compounds. They typically originate mostly from academia. Therefore, their R&D strategy is more focused on the elucidation of the biological roots of diseases rather than developing synthesis methods.

Agrochemicals

Agrochemical companies are the second largest users of fine chemicals. Most products have a "pharmaceutical heritage". As a consequence of an intensive M&A activity over the past 10–20 years, the industry now is more consolidated than the pharmaceutical industry. The top 10 companies, led by Syngenta, Switzerland; Bayer Cropsciences, Germany: Monsanto, USA; BASF Crop Protection, Germany, and Dow Agrosciences, USA have a share of almost 95% of the total 2'000'000 tons / $48.5 billion pesticide output in 2010. Since the 1990s the R&D effort is focused mainly on gene modified (GM) seeds. At both Monsanto and DuPont's seed subsidiary, Pioneer Hi-Bred, GM seed businesses already account for more than 50% of total sales. 100 new LMW agrochemicals have been launched in the period 2000–2009. However, only 8 products achieved sales in excess of \$100 million per year.

Generics play a bigger role in the agro than in the pharma industry. They represent some 70% of the global market. China National Chemical Corp, a.k.a. ChemChina Group, is the world's largest supplier of generic farm chemicals. Mahkteshim Agan, Israel, and Cheminova, Denmark follow on the ranks 2 and 3. Apart from these multibillion-dollar companies, there are hundreds of smaller firms with sales of less than \$50 million per year, mainly in India and China. The incidence of the cost of the active ingredient is about 33%; i.e., much higher than in drugs. Depending on the climatic conditions affecting crop yields, consumption and prices of agrochemicals are subject to wide fluctuations from year to year, impacting also the suppliers.

The molecular structures of modern agrochemicals are much more complex than in older products, but lower than of their pharma counterparts. The average molecular weight of the top 10 is 330, as compared with 477 for the top 10. In comparison to reagents used in pharmaceutical fine chemical syntheses, hazardous chemicals, e.g. sodium azide, halogens, methyl sulfide, phosgene, phosphorus chlorides, are more frequently used. Agrochemical companies sometimes outsource just these steps, which require specialized equipment, on toll conversion deals. With exception of the pyrethroids, which are photostable modifications of naturally occurring pyrethrums, active ingredients of agrochemicals rarely are chiral. Examples within *herbicides* are the world's longstanding top-selling product, Monsanto's round-up (glyphosate). Syngenta's cyclohexadione-type mesotrione and paraquat dichloride. Within *insecticides*, the traditional organophosphates, like malathion, and pyrethroids such as γ-cyhalotrin are being substituted for by neonicotinoids, like Bayer's imidacloprid and Syngenta's thiamethoxam and pyrazoles, such as BASF's fipronil. Chloranthaniliprole is the most important representative of Du Pont's award winning anthranilic diamide family of broad spectrum insecticides. Within *fungicides*, the strobilurins, a new class, are growing rapidly and already have captured more than 30% of the \$10 billion global fungicide market. Syngenta's azoxystrobin was the first product launched. Also BASF's F-500 Series, a.o. pyraclostrobin and kresoxim-methyl, Bayer CropScience, and Monsanto are developing new compounds in this class. Combination pesticides, such as Monsanto's Genuity and SmartStax are more and more frequently used.

Other Specialty Chemical Industries

Apart from life sciences, specialty chemicals -and therefore also their active ingredients, commodities or fine chemicals, as the case may be- are used ubiquitously, in both industrial applications, such as biocides and corrosion inhibitors in cooling water towers, and consumer applications, such as personal care and household products. The active ingredients extend from high-price / low-volume fine chemicals, used for liquid crystal displays to large-volume / low-price amino acids used as feed additives.

Industry	Sales ($ billion)	**Attrac-tiveness**	**Products**
Animal Health	~ 20	♦♦♦	Typical a.h. products derive from human medicines, e.g. Reconzile, dubbed "puppy Prozac". Paraciticides are the largest product category. Good growth prospects in fish farming.
Adhesives & Sealants	~ 60	♦♦	Uses extend from household, e.g. paper gluing, to high-tech specialty products for assembling electronic parts, automotive and aircraft construction.
Biocides	~ 3	♦	Largest applications are wood conversation and water treatment. A.I.'s mainly commodities
Catalysts & Enzymes	~ 15	♦	Catalysts (automotive, polymers, petroleum processing, chemicals) / enzymes (detergents / technical enzymes, food & feed) = 80 / 20
Dyestuffs & Pigments	~ 10	♦	Mostly based on large volume aromatic compounds, e.g., letter acids .Asian dyestuff prod, > 10^6 mtpa. Some niche products, e.g. colour shifting pigments
Electronic Chemicals	~ 30	♦♦♦	Sizeable and growing demand for fine chemicals, e.g. octafluorocyclobutane for etching. for liquid crystals and organic light emitting diodes (OLED).
Flavours & Fragrances	~ 20	♦♦	~3'000 molecules are used, e.g. (-) menthol [20'000 tons],polycyclic musks [10'000 tons], vanilline, linalool, geraniol, heterocyclics, 2-phenylethanol)
Food & Feed Additives	40-50	♦♦	Mainly amino acids (L-lysine [10^6 tons], L-methionine, ...), vitamins (C [> 10^5 tons], niacin, riboflavin, ...), artificial sweeteners (aspartame, splenda) and carotenoids
Specialty Polymers	NA	♦♦	Aerospace: Fluorinated poly ethylene/propylene,[30'000 tons], Polyether ether ketones [PEEK], Polyimides, Precision parts: Aramides [25'000 t], polybenzazoles

*fine chemicals merchant market size, growth potential

Examples of applications in eight areas, ranging from adhesives to specialty polymers, are listed in table . Overall, the attractiveness for the fine chemical industry is smaller than the life science industry. The total market, expressed in finished product sales, amounts to $150–200 billion, or about one fourth of the pharma market. The embedded fine chemicals account for an estimated $15 billion. Further disadvantages are the backward integration of the big players, e.g. Akzo-Nobel, Netherlands; Ajinomoto, Japan; Danone, France; Everlight Chemical Industrial Corp., Taiwan; Evonik-Degussa, Germany; Givaudan and Nestlé, Switzerland, Novozymes, Denmark, Procter & Gamble, and Unilever USA. Last but not least, innovation is rather based on new formulations of existing products, rather than the development of new fine chemicals. It is most likely to happen in application areas unrelated to human health (where NCEs are subject to very extensive testing).

Target Products and Services

Global sales of proprietary drugs are estimated $735 billion in 2010, or almost 90% of the total pharma market. Global sales of generics are about $100 billion, or just over 10% of the total pharma market. Due to the much lower unit price, their market share will be close to 30% on an API volume/volume basis.

Custom Manufacturing

The products and services offered by the fine chemical industry fall into two broad categories: (1) "Exclusives", a.k.a. custom manufacturing (CM) and (2) "standard" or "catalogue" products. "Exclusives", provided mostly under contract research or custom manufacturing arrangements, prevail in business with life science companies; "standards" prevail in other target markets. Service-intense custom manufacturing (CM) constitutes the most prominent activity of the fine chemical industry. CM is the antonym of outsourcing. In custom manufacturing, a specialty-chemicals company outsources the process development, pilot plant, and, finally, industrial-scale production of an active ingredient, or a predecessor thereof, to one, or a few, fine chemical companies. The intellectual property of the product, and generally also the manufacturing process, stay with the customer. The customer-supplier relationship is governed by an exclusive supply agreement. At the beginning of cooperation, the customer provides a "tech package," which in its simplest version, includes a laboratory synthesis description and SHE recommendations. In this case, the whole scale up, which comprises a factor of about one million (10 gram → 10 ton quantities), is done by the fine chemical company.

Standard products

Non-exclusives, "standard" or "catalogue products" constitute the second most important outlet for fine chemicals after custom manufacturing. API-for-Generics are the most important sub-category. Because of patent expiries, >60 "top 200 drugs" alone, representing aggregated sales of $140 billion, are expected to fall into the public domain within by 2015.and government incentives, global sales of generics are increasing rapidly. Asian companies dominate the API-for-Generics business. They have the triple advantage of their low cost basis, their big home markets, and previous manufacturing experience in producing for their domestic and other non-regulated markets.

Financials

Investment Costs

Investment costs for multipurpose plants are high in comparison with product output. However, they vary considerably, depending on the location, size of equipment and degree of sophistication (e.g., automation, containment, quality of equipment, complexity of infrastructure). An example for a cGMP multipurpose plant built in the US is shown in table below. The investment cost of $21 million comprises just the equipment and installation. The building, property and external services are excluded. For comparison purposes, the investment cost per m^3 reactor volume is used. In this case, it is $0.9 million. The amount includes the cost of the reaction vessel itself plus an equitable part of the ancillary equipment, like feeding tanks, piping, pumps & process control. If larger or smaller reactors were installed, the unit cost per m^3 would decrease or decrease with the exponent 0.5, respectively. Hence, by increasing the equipment size manufacturing costs on a per kilogram (kg^{-1}) basis typically decrease substantially. Also, costs for a plant that is used for the production of non regulated intermediates only would be substantially lower. Pharma companies tend to spend up to ten times more for a plant with the same capacity. In contrast, investment costs in developing countries, particularly in India or China, are considerably lower.

Equipment / Investment	Numbers
Description of main equipment	
Production trains	2
Reactor vessels (volume = 4 m^3)	6
…. Total reactor volume	…. 24 m^3
Filtration units	2
Dryers	2
Capital investment	
Total capital investment	**$21 million**
• Investment per production train	$ 11.5 million
• Investment per piece of main equipment	$ 2.1 million
• Investment per m3 reactor volume	$ 0.9 million

Manufacturing Costs

The raw material consumption and the conversion cost are the two elements that establish the manufacturing cost for a particular fine chemical. The former is determined primarily by the unit consumption and the purchasing cost of the materials used; the latter, by the throughput in kilograms per day in a given production bay. A precise calculation of the conversion cost is a demanding task. Different products with widely differing throughputs are produced in campaigns in multipurpose plants, occupying the equipment to different extents. Therefore, both the production capacity and the equipment utilization for a specific fine chemical are difficult to determine. Moreover, cost elements such as labor, capital, utilities, maintenance, waste disposal, and quality control cannot be allocated unambiguously.

An approximative calculation can be done by an experienced process development or pilot plant chemist on the basis of (1) the laboratory synthesis procedure and (2) by breaking down the process into unit operations, the standard costs of which have been determined previously Controlling has to be involved for a more in-depth costing.. The problems it has to address are how to fairly allocate costs for production capacity, which is not used. This can be due to the fact that part of a production bay is idle, because of lack of demand or because e.g., a reactor is not required for a particular process.

Manufacturing costs usually are reported on a per kilogram product basis. For the purpose of benchmarking (both internal and external), the volume x time / output (VTO), as mentioned above, is a useful aid.

Cost elements			Details	Share
raw materials			inclusive solvents	30 %
conversion cost	plant specific	utilities and energy	electric power, steam, brine	4-5 %
		plant labor	shift and daytime work	10-15 %
		capital cost	depreciation and interest on capital	15 %
		plant overhead	QC, maintenance, waste disposal, etc.	10 %
		Research & Development	inclusive pilot plant	8 %
	Marketing & Sales		inclusive promotion	5 %
	General Overhead		administrative services	15 %

An indicative cost structure for a fine chemical company is shown in table . Nowadays, a full 7-day/week operation, consisting of four or five shift teams, each working 8h per day, has become the standard. In terms of production costs, this is the most advantageous scheme. Higher salaries for night work are more than offset by better fixed cost absorption. As part of the budgeting process, standard costs for a production campaign of a particular fine chemical are determined on the basis of past experience. The actual results of the campaign are then compared with the standard. The capability of a fine chemical company to make dependable manufacturing cost forecasts is a distinct competitive advantage.

Profitability

The fine chemical industry has undergone several boom and bust phases during its almost 30 years of existence. The biggest boom took place in the late 1990s, when high-dosage, high volume anti-AIDS drugs and COX-2 inhibitors gave a big boost to custom manufacturing. After the end of the "irrational exuberance" in 2000, the industry suffered a first bust in 2003, as a result of capacity expansions, the advent of Asian competitors and a ruinous M&A activity, several billion dollars of shareholder value were destroyed. The most recent –minor- boom is associated with stockpiling of GlaxoSmithKline's Relenza (zanamivir) and Roche's Tamiflu (oseltamivir phosphate) by many countries in order to prepare for a possible avian flu epidemic. Surprisingly, the main cause for the 2009 slump had not been the general recession, but slow-downs of the growth and, even more so, inventory adjustments by the pharma industry. They resulted in postponements or cancellations of orders. The unfavorable development was in sharp contrast to the very optimistic growth forecasts, which many fine chemical companies, had announced. They had been based on equally promising sector reports from investment banks, which in turn had evolved from forward projections of the preceding boom period. In most cases, these projections have been missed by a large margin.

At the end of the "irrational exuberance" at the turn of the millennium and again in 2009 almost half of the industry achieved a return on sales (ROS) of more than 10%, and less than 10% an ROS below 5%. In the worst years, 2003 and 2009, almost half of the companies suffered from an ROS of less than 5%. Whereas during the period under review, 2000–2009. the average EBITDA / sales and EBIT / sales ratios of representative companies, resp. divisions were 15% and 7½%, respectively, in the period 2000–2009, the numbers were 20% and 10–13% in the boom, and 10% and 5% in the bust phases. The factor 2 between the high and low numbers reflects the volatility of the industry's profitability. All in all, the average Western fine-chemical firms have been making a return below the cost of capital, i.e. they are not reinvestment grade.

Outlook

Two main trends impinge on the industry. On the *supply side*, biotechnology is rapidly gaining importance. In the synthesis of small molecule fine chemicals, the use of biocatalysts and microbial fermentation enable both a more sustainable and economic production than conventional organic chemistry. In the synthesis of big molecules, such as biopharmaceuticals, it is the method of choice. Biopharmaceuticals are expected grow 15% per year, three times as fast as small molecule drugs. Five of the top ten drugs were biopharmaceuticals in 2010, and this is expected to grow to eight by 2016.

On the *demand side*, the main customer base for fine chemicals, the pharmaceutical industry, is faced with slower growth of demand, patent expirations of many lucrative blockbuster drugs and stalling new product launches. In order to restrain these challenges, the leading companies are implementing restructuring programs. They comprise a reduction of in-house chemical manufacturing and plant eliminations. Outsourcing is moving up from a purely opportunistic to a strategic approach. It is difficult to make a judgment, whether the positive or negative effects of these initiatives will prevail. In a worst-case scenario, a condition could develop, whereby even top-tier mid-sized, family-owned fine-chemical companies with state-of-the-art plants and processes could be relegated to producing small quantities of fine chemicals for new life-science products in late stage of development. In agro fine chemicals, the active ingredients become more sophisticated and performing. Therefore, they require multipurpose instead of dedicated plants prevailing in the industry so far. At the same token, outsourcing is gaining ground.

Globalization results in a shift of fine chemical production from the industrialized to developing countries. The latter benefit not only from a "low cost / high skill" advantage, but also from a rapidly rising domestic demand for Western medicine. Despite the mantras of Western industry leaders, the cost advantage of the Asian producers is going to persist. As the pharmemerging countries mainly use generics, their market share continues to grow to the detriment of originator pharmaceuticals and agrochemicals. This is also the case for biosimilars, the generic versions of biopharmaceuticals. As a consequence of the harsh business climate, many Western fine chemical companies or divisions created during the "irrational exuberance" at the end of the 20th century already have exited from the sector. Others will follow suit or will be acquired by private equity firms. Survival strategies include implementation of lean production principles originally developed by the automotive industry and extending the business model to include also contract research at the beginning and active drug formulation towards the end of the added value chain. This latter strategy, however, is not finding unanimous approval by industry experts.

Although the demand for fine chemicals on the merchant market has not grown to the extent originally anticipated, fine chemicals still provide attractive opportunities for well-run companies, which are fostering the critical success factors, namely running fine chemicals as a core business, pursuing niche technologies—primarily biotechnology—and taking advantage of the opportunities offered by the Asian market.

COMMODITY CHEMICALS

Commodity chemicals (or bulk commodities or bulk chemicals) are a group of chemicals that are made on a very large scale to satisfy global markets. The average prices of commodity chemicals are regularly published in the chemical trade magazines and web sites such as Chemical Week and ICIS. There have been several studies of the scale and complexity of this market for example in the USA.

Commodity Chemicals is a sub-sector of the chemical industry (other sub sectors are fine chemicals, speciality chemicals, inorganic chemicals, petrochemicals, pharmaceuticals, renewable energy (e.g. biofuels) and materials (e.g. biopolymers)) Commodity Chemicals are differentiated by primarily their bulk of their manufacture.

Industrial Organic Chemistry

The chemistry and processes used to create organic commodity chemicals (that is those based on the chemistry of carbon) from basic raw materials such as oil and gas; plus the chemistry and processes that are used to manufacture the many downstream speciality and fine chemicals have been studied in universities and industrial research centres for most of the 20th century. This is called organic synthesis and the industrial process streams that have been created are termed Industrial Organic Chemistry. and these can be found in many organic chemistry text books.

Types

Chemical compounds are often classified into two classes, inorganic and organic. Both are produced on large scale.

List

Commonly traded commodity chemicals include:

- Acetic acid
- Acetone
- Acrylate esters
- Acrylic acid
- Acrylonitrile
- Adipic acid
- Benzene

- Biodiesel
- Bisphenol A
- Butadiene
- Butanediol
- Butyl acetate
- Castor oil
- Caprolactam
- Cumene
- Cyclohexane
- Dimethyl terephthalate
- Epichlorohydrin
- Epoxy resins
- ETBE
- Ethanol
- Ethanolamines
- Ethyl acetate
- Ethylene
- Ethylene diamine
- Ethylene dichloride (EDC)
- Ethylene oxide
- Ethylene vinyl acetate
- Ethyl benzene
- Expandable polystyrene
- Fatty acids
- Fatty alcohols
- Glycerine
- Glycol ethers
- Hexane
- Isocyanates
- Maleic anhydride
- Melamine
- Methanol
- Methyl isobutyl ketone (MIBK)
- Methyl ethyl ketone (MEK)
- Methyl methacrylate
- Monoethylene glycol

- Monopropylene glycol
- Naphtha
- Nylon 6 & 6/6,
- Oxo alcohols
- Perchloroethylene
- Phenol
- Phthalic anhydride
- Polycarbonate
- Polyethylene
- Polyethylene terephthalate (PET)
- Polymethyl methacrylate (PMMA)
- Polyols
- Polypropylene
- Polystyrene
- Polyvinyl chloride
- Propylene
- Propylene oxide
- Purified terephthalic acid (PTA)
- Recycled polyethylene terephthalate (PET)
- Sorbitol
- Styrene
- Toluene
- Urea
- Vinyl acetate
- Vinyl chloride
- White spirit
- Xylenes

Manufacture

Materials such as these are made in large scale chemical manufacturing locations around the world. Not all of the materials are produced in one single location but groups of related materials often are to induce industrial symbiosis as well as material, energy and utility efficiency and other economies of scale. These locations often have clusters of manufacturing units called chemical plants that share utilities and large scale infrastructure such as power stations, port facilities, road and rail terminals. In the United Kingdom for example there are 4 main locations for commodity chemical manufacture: near the River Mersey in Northwest England, on the Humber on the East coast of Yorkshire, in Grangemouth near the Firth of Forth in Scotland and on Teesside as part of the Northeast of

England Process Industry Cluster (NEPIC). Approximately 50% of the UK's petrochemicals, which are also commodity chemicals, are produced by the industry cluster companies on Teesside on three large chemical parks at Wilton, Billingham and Seal Sands.

These useful materials are mostly manufactured in continuous process chemical plants, as opposed to batch manufacturing chemical plants which are commoner in the other aforementioned sub-sectors of the industry, particularly speciality chemicals and fine chemicals. The object in commodity chemical manufacture is to carry out the chemical transformations on the raw materials used on a massive scale, producing relatively simple molecules at the lowest possible cost. Investment into these large scale manufacturing units and the lowering of environmental impacts in commodity chemical production involves large capital expenditures and long time horizons. These are known as capital intensive industries.

Research Trends

Research has been conducted on the use of alternative feedstocks such as biomass and societal waste. These lower carbon processes are sometimes termed "Green Chemistry".

SPECIALITY CHEMICALS

Speciality chemicals (also called specialties or effect chemicals) are particular chemical products which provide a wide variety of effects on which many other industry sectors rely. Some of the categories of speciality chemicals are adhesives, agrichemicals, cleaning materials, cosmetic additives, construction chemicals, elastomers, flavours, food additives, fragrances, Industrial gases, lubricants, polymers, surfactants, and textile auxiliaries. Other industrial sectors such as automobile, aerospace, food, cosmetics, agriculture, manufacturing, textile industries are highly dependent on such products.

Speciality chemicals are materials used on the basis of their performance or function. Consequently, in addition to "effect" chemicals they are sometimes referred to as "performance" chemicals or "formulation" chemicals. They can be unique molecules or mixtures of molecules known as formulations. The physical and chemical characteristics of the single molecules or the formulated mixtures of molecules and the composition of the mixtures influences the performance end product. In commercial applications the companies providing these products more often than not provide targeted customer service to innovative individual technical solutions for their customers. This is a differentiating component of the service provided by speciality chemical producers when they are compared to the other sub-sectors of the chemical industry such as fine chemicals, commodity chemicals, petrochemicals and pharmaceuticals.

In the USA the speciality chemical manufacturers are members of the Society of Chemical Manufacturers and Affiliates (SOCMA). In the United Kingdom such companies are members of the British Association of Chemical Specialties (BACS). SOCMA state that "Specialty chemicals differ from commodity chemicals in that each one may have only one or two uses, while commodities may have dozens of different applications for each chemical. While commodity chemicals make up most of the production volume (by weight) in the global marketplace, specialty chemicals make up most of the diversity (number of different chemicals) in commerce at any given time."

Manufacturing Speciality Chemicals

Speciality chemicals are usually manufactured in batch chemical plants using batch processing techniques. A batch process is one in which a defined quantity of product is made from a fixed input of raw materials during a measured period of time. The batch process most often consists of introducing accurately measured amounts of starting materials into a vessel followed by a series of processes involving mixing, heating & cooling, making more chemical reactions, distillation, crystallization, separation, drying, packaging etc., taking place at predetermined and scheduled intervals. The manufacturing processes are supported by activities such as the quality testing, storage& warehousing, logistics of the products, and management by recycling, treatment and disposal of by-products, and waste streams. For the next "batch" the equipment may be cleaned and the above processes repeated.

Most speciality chemicals are organic chemicals that are used in a wide range of every day products used by consumers and industry. It is a consumer driven sector and as such the specialty chemical industry has to be innovative, entrepreneurial and consumer-driven. In contrast to the production of commodity chemicals that are usually made on large scale single product manufacturing units to achieve economies of scale, specialty manufacturing units are required to be flexible because the products, raw materials, processes &, operating conditions and equipment mix may change on a regular basis to respond to the needs of customers.

In the United Kingdom there are a many speciality chemical companies that are members of the British Association of Chemical Specialties (BACS) and also the Chemical Industries Association (CIA). Most of these companies have their manufacturing units based in the North of England for example see the membership of the Northeast of England they are members of the Northeast of England process Industry Cluster (NEPIC).

The Global Speciality Chemical Market

In 2011 the Global Specialty Chemicals market had total revenues of $767.5 billion and it is reported that it had a compound annual growth rate (CAGR) of 2.7% between 2007 and 2011. The market is further projected to reach $980 billion by the end of 2016. These speciality products are marketed as pesticides, speciality polymers, electronic chemicals, surfactants, construction chemicals, Industrial Cleaners, flavors and fragrances, speciality coatings, printing inks, water-soluble polymers, food additives, paper chemicals, oil field chemicals, plastic adhesives, adhesives and sealants, cosmetic chemicals, water management chemicals, catalysts, textile chemicals.

The world's top five specialty chemicals segments in 2012 were specialty polymers, industrial and institutional (I&I) cleaners, construction chemicals, electronic chemicals, and flavors and fragrances. These segments had a market share of about 36% The ten largest segments accounted for 62% of total annual specialty chemicals sales.

Speciality Chemical Companies

The speciality chemical market is complex and each specialty chemicals business segment comprises many sub-segments, each with individualized product, market and competitive profiles. This has given rise to a wide range of business needs and opportunities, consequently there are a large number of speciality chemical companies around the world. Many of these companies are SME's with their own niche products and sometimes technology focus. The common stock of over 400 speciality chemical companies from around the world are identified by Bloomberg the providers of global business and financial information. There are many more privately owned speciality chemical companies that are not quoted on the global stock markets.

The 10 largest European speciality chemical companies are BASF, AkzoNobel, Clariant, Evonik, Cognis, Kemira, Lanxess, Rhodia, Wacker and Croda. By definition, speciality chemicals are produced in relatively small quantities but they represent 28 per cent of EU chemical sales.

The 10 largest USA Speciality Chemical Companies are Huntsman, Ashland, Chemtura, Rockwood, Albemarle, Cabot, W. R. Grace, Ferro Corporation, Cytec Industries and Lubrizol.

The emergence of India as a manufacturer and supplier of speciality chemicals has had a major impact on the Global speciality chemical industry. In India the many speciality chemical companies are members of national organisations such as the Indian Chemical Council (ICC) and the Indian speciality chemical manufacturers association (ISMCA). The wide capability of these companies extends into all sectors and sub-sectors of the speciality chemical market.

The United Kingdom has 1300 speciality chemical companies and which have an annual turnover of £11.2 billion. The products of these UK companies are sold globally and contribute significantly to the UKs export trade. With over £30 billion of exports the chemical industry is the last remaining net-exporting manufacturing industry in the UK and Speciality Chemicals make up a significant proportion of this. The products include dyestuffs, paints, explosives, adhesives, flavors and fragrances, photographic chemicals, unrecorded media and various industrial specialities. As Speciality Chemical manufacturers, unlike commodity chemical manufacturers, are less dependent on large scale infrastructure, therefore Speciality Chemical companies can be found in almost all regions of the UK. Some 80% of the United Kingdom Chemical industry is based in the north of the country and consequently there are concentrations of Speciality Chemical companies in Yorkshire, and in the membership of the Northeast of England Process Industry Cluster (NEPIC).

PETROCHEMICAL

Petrochemicals, also called petroleum distillates, are chemical products derived from petroleum. Some chemical compounds made from petroleum are also obtained from other fossil fuels, such as coal or natural gas, or renewable sources such as corn or sugar cane.

The two most common petrochemical classes are olefins (including ethylene and propylene) and aromatics (including benzene, toluene and xylene isomers). Oil refineries produce olefins and aromatics by fluid catalytic cracking of petroleum fractions. Chemical plants produce olefins by steam cracking of natural gas liquids like ethane and propane. Aromatics are produced by catalytic reforming of naphtha. Olefins and aromatics are the building-blocks for a wide range of materials such as solvents, detergents, and adhesives. Olefins are the basis for polymers and oligomers used in plastics, resins, fibers, elastomers, lubricants, and gels.

Petrochemical plant in the Kingdom of Saudi Arabia.

Global ethylene and propylene production are about 115 million tonnes and 70 million tonnes per annum, respectively. Aromatics production is approximately 70 million tonnes. The largest petrochemical industries are located in the USA and Western Europe; however, major growth in new production capacity is in the Middle East and Asia. There is substantial inter-regional petrochemical trade.

Primary petrochemicals are divided into three groups depending on their chemical structure:

- Olefins includes ethylene, propylene, and butadiene. Ethylene and propylene are important sources of industrial chemicals and plasticsproducts. Butadiene is used in making synthetic rubber.
- Aromatics includes benzene, toluene, and xylenes. Benzene is a raw material for dyes and synthetic detergents, and benzene and toluene for isocyanates MDI and TDI used in making polyurethanes. Manufacturers use xylenes to produce plastics and synthetic fibers.
- Synthesis gas is a mixture of carbon monoxide and hydrogen used to make ammonia and methanol. Ammonia is used to make the fertilizerurea and methanol is used as a solvent and chemical intermediate.

The prefix "petro-" is an arbitrary abbreviation of the word "petroleum"; since "petro-" is Ancient Greek for "rock" and "oleum" means "oil". Therefore, the etymologically correct term would be "oleochemicals". However, the term oleochemical is used to describe chemicals derived from plant and animal fats.

Sources

The adjacent diagram schematically depicts the major hydrocarbon sources used in producing petrochemicals are:

- Methane, ethane, propane and butanes: Obtained primarily from natural gas processing plants.
- Naphtha obtained from petroleum refineries.
- Benzene, toluene and xylenes, as a whole referred to as BTX and primarily obtained from petroleum refineries by extraction from the reformate produced in catalytic reformers.
- Gas obtained from petroleum refineries.

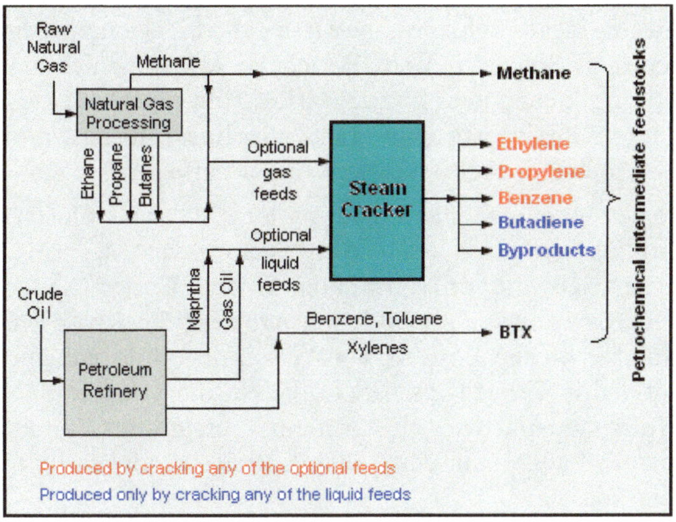

Petrochemical feedstock sources

Methane and BTX are used directly as feedstocks for producing petrochemicals. However, the ethane, propane, butanes, naphtha and gas oil serve as optional feedstocks for steam-assisted thermal cracking plants referred to as *steam crackers* that produce these intermediate petrochemical feedstocks:

- Ethylene
- Propylene
- Butenes and butadiene
- Benzene

In 2007, the amounts of ethylene and propylene produced in steam crackers were about 115 Mt (megatonnes) and 70 Mt, respectively. The output ethylene capacity of large steam crackers ranged up to as much as 1.0 – 1.5 Mt per year.

Steam crackers are not to be confused with steam reforming plants used to produce hydrogen and ammonia.

Manufacturing Locations

Like commodity chemicals, petrochemicals are made on a very large scale. Petrochemical manufacturing units differ from commodity chemical plants in that they often produce a number of related products. Compare this with specialty chemical and fine chemical manufacture where products are made in discrete batch processes.

Petrochemicals are predominantly made in a few manufacturing locations around the world, for example in Jubail & Yanbu Industrail Cities in Sauid Arabia, Texas & Louisiana in the USA, in Teesside in the Northeast of England in the United Kingdom, in Rotterdam in the Netherlands, and in Jamnagar & Dahej in Gujarat, India. Not all of the petrochemical or commodity chemical materials produced by the chemical industry are made in one single location but groups of related materials are often made in adjacent manufacturing plants to induce industrial symbiosis as well as material and utility efficiency and other economies of scale. This is known in chemical engineering terminology as integrated manufacturing. Speciality and fine chemical companies are sometimes found in similar manufacturing locations as petrochemicals but, in most cases, they do not need the same level of large scale infrastructure (e.g., pipelines, storage, ports and power, etc.) and therefore can be found in multi-sector business parks.

The large scale petrochemical manufacturing locations have clusters of manufacturing units that share utilities and large scale infrastructure such as power stations, storage tanks, port facilities, road and rail terminals. In the United Kingdom for example, there are 4 main locations for such manufacturing: near the River Mersey in Northwest England, on the Humber on the East coast of Yorkshire, in Grangemouth near the Firth of Forth in Scotland and in Teesside as part of the Northeast of England Process Industry Cluster (NEPIC). To demonstrate the clustering and integration, some 50% of the United Kingdom's petrochemical and commodity chemicals are produced by the NEPIC industry cluster companies in Teesside.

List of Significant Petrochemicals and Their Derivatives

The following is a partial list of the major commercial petrochemicals and their derivatives:

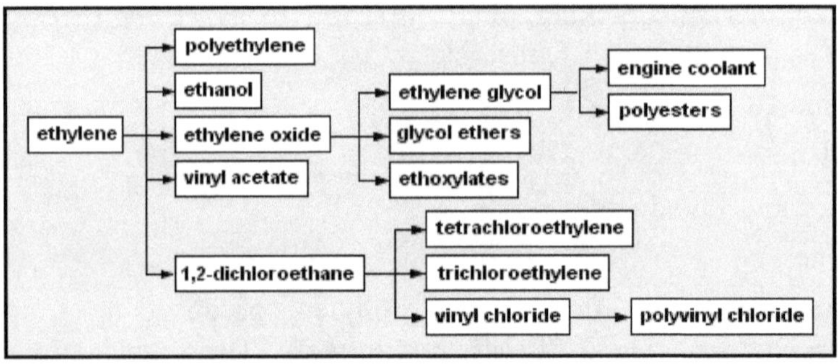

Chemicals produced from ethylene

- ethylene - the simplest olefin; used as a chemical feedstock and ripening stimulant
  - polyethylene - polymerized ethylene
  - ethanol - via ethylene hydration (chemical reaction adding water) of ethylene
  - ethylene oxide - via ethylene oxidation
    - ethylene glycol - via ethylene oxide hydration
      - engine coolant - ethylene glycol, water and inhibitor mixture
      - polyesters - any of several polymers with ester linkages in the main chain
    - glycol ethers - via glycol condescension
    - ethoxylates
  - vinyl acetate
  - 1,2-dichloroethane
    - trichloroethylene
    - tetrachloroethylene - also called perchloroethylene; used as a dry cleaning solvent and degreaser
    - vinyl chloride - monomer for polyvinyl chloride
      - polyvinyl chloride (PVC) - type of plastic used for piping, tubing, other things

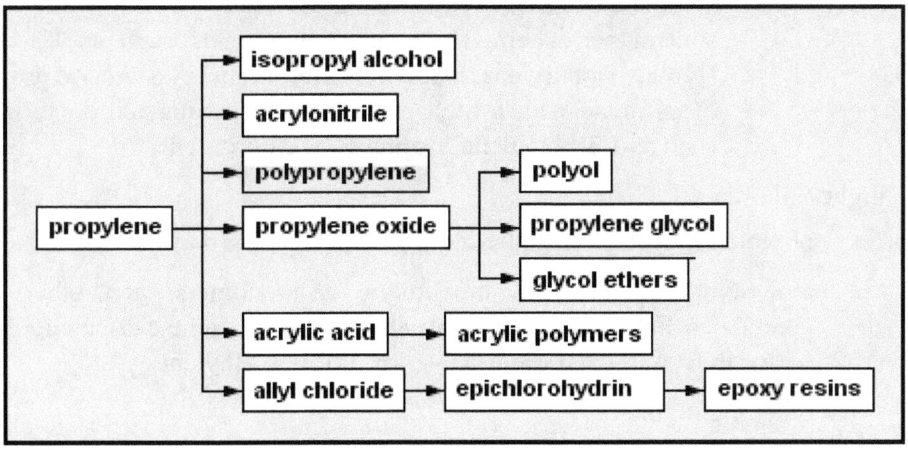

Chemicals produced from propylene

- propylene - used as a monomer and a chemical feedstock
  - isopropyl alcohol - 2-propanol; often used as a solvent or rubbing alcohol
  - acrylonitrile - useful as a monomer in forming Orlon, ABS

- polypropylene - polymerized propylene
- propylene oxide
  - polyether polyol - used in the production of polyurethanes
  - propylene glycol - used in engine coolant and aircraft deicer fluid
  - glycol ethers - from condensation of glycols
- acrylic acid
  - acrylic polymers
- allyl chloride -
  - epichlorohydrin - chloro-oxirane; used in epoxy resin formation
    - epoxy resins - a type of polymerizing glue from bisphenol A, epichlorohydrin, and some amine

■ C4 hydrocarbons - a mixture consisting of butanes, butylenes and butadienes

- isomers of butylene - useful as monomers or co-monomers
  - isobutylene - feed for making methyl *tert*-butyl ether (MTBE) or monomer for copolymerization with a low percentage of isoprene to make butyl rubber
- 1,3-butadiene (or buta-1,3-diene) - a diene often used as a monomer or co-monomer for polymerization to elastomers such as polybutadiene, styrene-butadiene rubber, or a plastic such as acrylonitrile-butadiene-styrene (ABS)
  - synthetic rubbers - synthetic elastomers made of any one or more of several petrochemical (usually) monomers such as 1,3-butadiene, styrene, isobutylene, isoprene, chloroprene; elastomeric polymers are often made with a high percentage of conjugated diene monomers such as 1,3-butadiene, isoprene, or chloroprene

■ higher olefins

- polyolefins such poly-alpha-olefins, which are used as lubricants
- alpha-olefins - used as monomers, co-monomers, and other chemical precursors. For example, a small amount of 1-hexene can be copolymerized with ethylene into a more flexible form of polyethylene.
- other higher olefins
- detergent alcohols

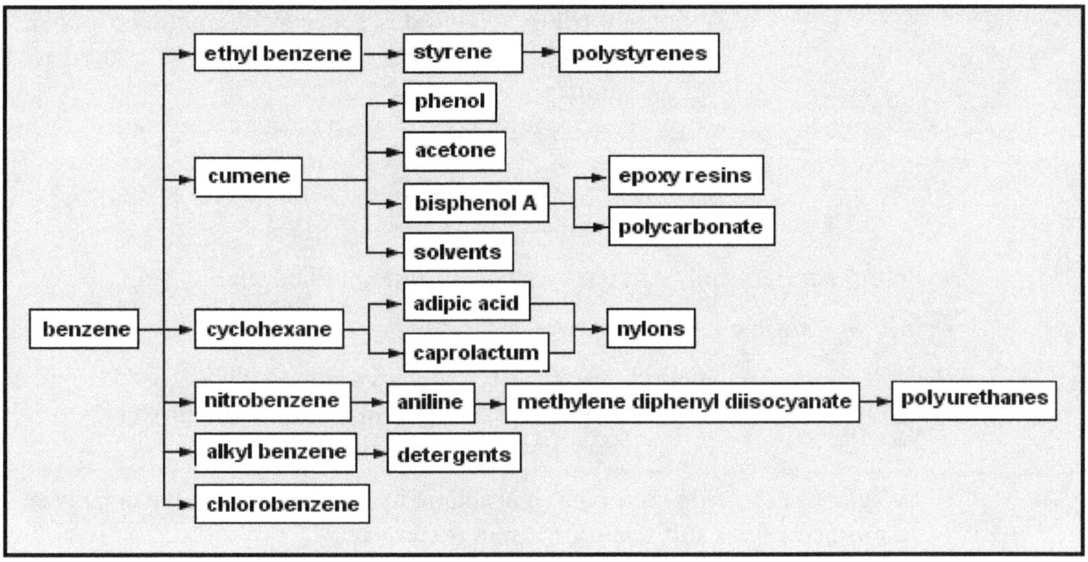

Chemicals produced from benzene

- benzene - the simplest aromatic hydrocarbon
  - ethylbenzene - made from benzene and ethylene
    - styrene made by dehydrogenation of ethylbenzene; used as a monomer
      - polystyrenes - polymers with styrene as a monomer
  - cumene - isopropylbenzene; a feedstock in the cumene process
    - phenol - hydroxybenzene; often made by the cumene process
    - acetone - dimethyl ketone; also often made by the cumene process
    - bisphenol A - a type of "double" phenol used in polymerization in epoxy resins and making a common type of polycarbonate
      - epoxy resins - a type of polymerizing glue from bisphenol A, epichlorohydrin, and some amine
      - polycarbonate - a plastic polymer made from bisphenol A and phosgene (carbonyl dichloride)
    - solvents - liquids used for dissolving materials; examples often made from petrochemicals include ethanol, isopropyl alcohol, acetone, benzene, toluene, xylenes
  - cyclohexane - a 6-carbon aliphatic cyclic hydrocarbon sometimes used as a non-polar solvent
    - adipic acid - a 6-carbon dicarboxylic acid, which can be a precursor used as a co-monomer together with a diamine to form an alternating copolymer form of nylon.

        - nylons - types of polyamides, some are alternating copolymers formed from copolymerizing dicarboxylic acid or derivatives with diamines
    - caprolactam - a 6-carbon cyclic amide
        - nylons - types of polyamides, some are from polymerizing caprolactam
- nitrobenzene - can be made by single nitration of benzene
    - aniline - aminobenzene
        - methylene diphenyl diisocyanate (MDI) - used as a co-monomer with diols or polyols to form polyurethanes or with di- or polyamines to form polyureas
- alkylbenzene - a general type of aromatic hydrocarbon, which can be used as a presursor for a sulfonate surfactant (detergent)
    - detergents - often include surfactants types such as alkylbenzenesulfonates and nonylphenol ethoxylates
- chlorobenzene

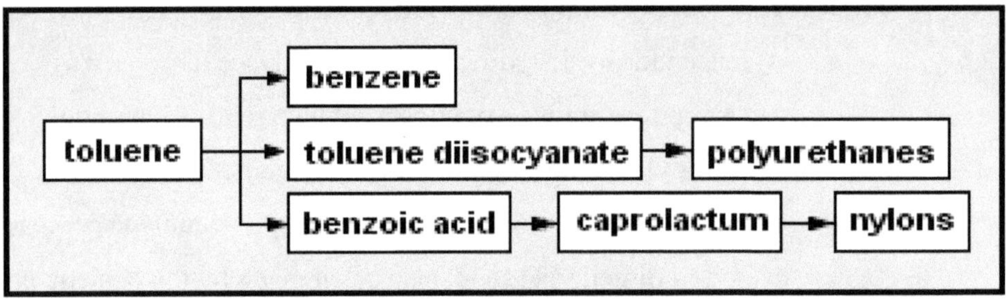

Chemicals produced from toluene

■ toluene - methylbenzene; can be a solvent or precursor for other chemicals

- benzene
- toluene diisocyanate (TDI) - used as co-monomers with polyether polyols to form polyurethanes or with di- or polyamines to form polyureas polyurethanes
- benzoic acid - carboxybenzene
    - caprolactam
        - nylon

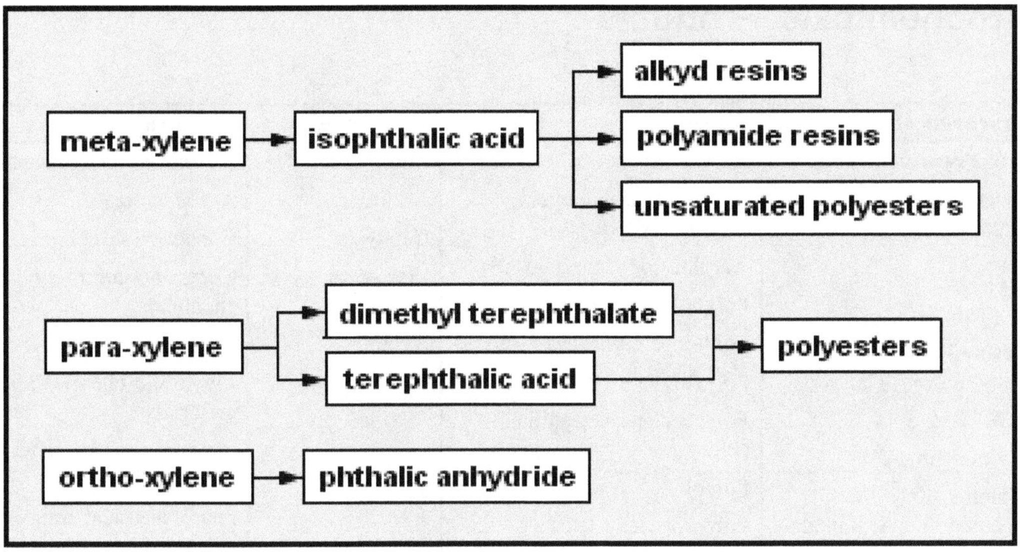

Chemicals produced from xylenes

- mixed xylenes - any of three dimethylbenzene isomers, could be a solvent but more often precursor chemicals
  - *ortho*-xylene - both methyl groups can be oxidized to form (*ortho*-)phthalic acid
    - phthalic anhydride
  - *para*-xylene - both methyl groups can be oxidized to form terephthalic acid
    - dimethyl terephthalate - can be copolymerized to form certain polyesters
      - polyesters - although there can be many types, polyethylene terephthalate is made from petrochemical products and is very widely used.
    - purified terephthalic acid - often copolymerized to form polyethylene terephthalate
      - polyesters
  - *meta*-xylene
    - isophthalic acid
      - alkyd resins
      - Polyamide Resins
      - Unsaturated Polyesters

Petrochemicals Products

Petrochemicals	Fibers	Petroleum	Chemicals
Basic Feedstock Benzene Butadiene Ethylene *p*-Xylene Propylene **Intermediates** 2-Ethylhexanol (2-EH) Acetic acid Acrylonitrile (AN) Ammonia Bis(2-ethylhexyl) phthalate (dioctyl phthalate) n-Butene Cyclohexane Dimethyl terephthalate (DMT) Dodecylbenzene Ethanol Ethanolamine Ethoxylate 1,2-Dichloroethane (ethylene dichloride or EDC) Ethylene glycol (EG) Ethylene oxide (EO) Formaldehyde Moulding Compound (FMC) n-Hexene Linear alkyl benzene (LAB) Methanol Methyl tert-butyl ether (MTBE) Phenol Propylene oxide Purified terephthalic acid (PTA) Styrene monomer (SM) Thermosetting Resin (Urea/Melamine) Vinyl acetate monomer (VAM) Vinyl chloride monomer (VCM)	Acrylic fiber Acrylonitrile butadiene styrene (ABS) Acrylonitrile styrene (AS) Polybutadiene (PBR) Polyvinyl chloride (PVC) Polyethylene (PE) Polyethylene terephthalate (PET) Polyol Polypropylene (PP) Polystyrene (PS) Styrene butadiene (SBR) Acrylic-formaldehude (AF)	Lubricants Additives Catalysts Marine fuel oil Petroleum refining	Adhesives and sealants Agrochemicals Construction chemicals Corrosion control chemicals Cosmetics raw materials Electronic chemicals and materials Flavourings, fragrances, food additives Pharmaceutical drugs Specialty and industrial chemicals Specialty and industrial gases Inks, dyes and printing supplies Packaging, bottles, and containers Paint, coatings, and resins Polymer additives Specialty and life sciences chemicals Surfactants and cleaning agents

REFERENCES

1. Ellison-Taylor; et al. (1970). Chemical Plant Technology: An Introductory Manual. Longmans.
2. Douglas, James M. (1988). Conceptual Design of Chemical Processes. McGraw-Hill. ISBN 0-07-017762-7.
3. Stork, William (2004). "Speciality Chemicals" (PDF). Chemical & Engineering News supplement 82. pp. 35–39.
4. Plant Construction Location Factor - Intratec Knowledge Base Archived 6 March 2014 at the Wayback Machine.
5. Delgass; et al. "Seventy Five Years of Chemical Engineering". Prudue University. Retrieved 13 August 2013.
6. Cussler, Moggridge and Moggridge (2001). Chemical product design. Cambridge University Press.
7. UK Trade&Investment. "Chemicals–the UK advantage" (PDF). p. 9-10. Retrieved 10 July 2013.
8. Hurworth, Colin (1999). Wilton the First Fifty Years. Falcon Press. ISBN 1872339018.
9. "Chemical Accidents: About Archived May 31, 2008, at the Wayback Machine.." OECD Environment Directorate. OECD. 19 July 2007.
10. Jones, Dr. Ron. "UK Chemical Reaction Hazard Forum Home Page". Retrieved 22 October 2016.
11. Pintelon, L. & Puyvelde, F. V., 1997. Estimating Plant Construction Costs. Chemical Engineering, August, pp. 98-104
12. Humphreys, K. K., 2005. Project and Cost Engineers' Handbook. 4th ed. s.l.:Marcel Dekker
13. Peters, M., Timmerhaus, K. & West, R., 2004. Plant Design and Economics for Chemical Engineers. 5th ed. s.l.:McGraw-Hill
14. Vatavuk, W. M., 2002. Updating the CE Plant Cost Index. Chemical Engineering, January, pp. 62-70
15. Couper, J. R., 2003. Process Engineering Economics. s.l.:Marcel Dekker
16. Albright, L.. 2008. Albright's Chemical Engineering Handbook. s.l.:CRC-Press
17. http://www.chemengonline.com/pci/
18. http://www.equipment-cost-index.com/
19. "Archived copy". Archived from the original on 2012-09-29. Retrieved 2012-05-08.
20. http://www.intratec.us/ic
21. "Plant Cost Indices". Process Engineering. 93 (3): 62. May 2012.
22. A.Kleemann; J. Engel; B. Kutscher; D. Reichert (2009). Pharmaceutical Substances (5th ed.). pp. 291–292.

23. A.Kleemann; J. Engel; B. Kutscher; D. Reichert (2009). Pharmaceutical Substances (5th ed.). pp. 1189–1191.
24. E. Reeder; L. H. Sternbach (1968 to Hoffmann-LaRoche). US 3371085. Check date values in: |date= (help)
25. Hughes, Andrew B. (2011). Amino Acids, Peptides and Proteins in Organic Chemistry. Volumes 1-5: John Wiley & Sons, Hoboken.
26. D. Bellus, S. V. Ley, R. Noyori et al. (Series editors) (2010). Science of Synthesis: Houben-Weyl Methods of Molecular Transformations. Thieme Verlag, Stuttgart.
27. Examples: Swiss Federal Institute of Technology (ETHZ), Switzerland; Massachusetts Institute of Technology (MIT), USA; Institut für Mikrotechnik (IMM), Germany; University of Washington (WU), USA; Micro-Chemical ProcessTechnology Research Association (MCPT), Japan.
28. V. Hessel; A. Renken; J.C. Shouten; J. Yoshida (2009). Micro Process Engineering. Wiley-VCH Verlag, Weiheim.
29. Wim Soetaert; Erick J. Vandamme (2010).), Industrial Biotechnology: Sustainable Growth and Economic Success. J. Wiley & Sons, Hoboken NJ.
30. Secretary: Prof. M.P. Walsh. University of Calgary, Calgary, Canada T2N 4N1
31. Victor A. Vinci; Sarad R. Parekh (2010). Handbook of Industrial Cell Culture: Mammalian, Microbial, and Plant Cells. Humana Press, New York.
32. A. Chassin; P. Pollak (January–February 2004). "Outlook for chemical and biochemical manufacturing, PharmaChem". 1–2: 23–26.
33. "Expected Top Ten Drugs by 2016". Retrieved 11 December 2011.
34. A.Gosh; S. Ray; G. Jain; A. Arora (2011). CRAMS India: Overview & Outlook. ICRA Ltd. Mumbai.
35. Samuel L. Tuthill, Norman C. Jamieson, Kirk-Othmer (2000). Encyclopedia of Chemical Technology (4th ed.). John Wiley & Sons, Hoboken NJ. p. 857.
36. The U.S. Pharmacopeia 34 (USP 34 –NF29). The U.S Pharmacopeial Convention, Inc., Rockville, MD. 2011.
37. Adapted from: Pollak, Peter (2011). Fine Chemicals – The Industry and the Business, (2nd. rev. ed.). Table 2.2: J. Wiley & Sons. p. 1. ISBN 978-0-470-62767-9.
38. CPhI Worldwide, 25-27 October 2011. Messe Frankfurt, UBM plc., London.
39. Stanley H. Nusim (2009). Active Pharmaceutical Ingredients: Development, Manufacturing, and Regulation (2nd ed.). Taylor&Francis Group, Boca Ratpn FL. pp. 9–91.
40. Pollak, Peter (2011). Fine Chemicals – The Industry and the Business (2nd. rev. ed.). Table 6.1: J. Wiley & Sons. p. 69. ISBN 978-0-470-62767-9.
41. A. Aboody; B. Lev (2001). R&D Productivity in the Chemical Industry. New York University, Stern School of Business.
42. B. Lev (Winter 1999). "Journal of Applied Corporate Finance". 21–35.

43. Pollak, Peter (2011). Fine Chemicals – The Industry and the Business (2nd. rev. ed.). Table 10.2: J. Wiley & Sons. p. 105. ISBN 978-0-470-62767-9.

44. C.D.S. Tomlin (2011). The Pesticide Manual: A World Compendium (15th ed.). BCPC Publications, Alton, Hampshire, UK.

45. Pollak, Peter (2011). Fine Chemicals – The Industry and the Business (2nd. rev. ed.). Table 7.1: J. Wiley & Sons. p. 76. ISBN 978-0-470-62767-9.

46. Pollak, Peter (2011). Fine Chemicals – The Industry and the Business (2nd. rev. ed.). Table 7.3: J. Wiley & Sons. p. 79. ISBN 978-0-470-62767-9.

47. Guy Villax (Nov–Dec 2008). "Family owned businesses". Chimica Oggi / chemistry today. 26 (6): 8.

48. P. Pollak; A. Badrot; R. Dach; A. Swadi (Nov–Dec 2011). Costs of Asian Fine Chemical Producers close-up to European Levels – Facts or Fiction?. Contract Pharma.

CHAPTER 9

Chemical Thermodynamics

Chapter Contents

- **Introduction**
- **History**
- **Overview**
- **Chemical Energy**
- **Non Equilibrium**

Chemical thermodynamics is the study of how heat and work relate to each other both in changes of state and in chemical reactions. It involves a series of rules and laws that explain how heat and work well, work, and explains which processes can happen spontaneously and which need some help.

There are several basic principles of chemical thermodynamics to consider: systems, the laws of thermodynamics, and enthalpy. Chemical thermodynamics is also concerned with four particular quantities: internal energy, enthalpy, entropy and the Gibbs free energy.

INTRODUCTION

Chemical thermodynamics is the study of the interrelation of heat and work with chemical reactions or with physical changes of state within the confines of the laws of thermodynamics. Chemical thermodynamics involves not only laboratory measurements of various thermodynamic properties, but also the application of mathematical methods to the study of chemical questions and the *spontaneity* of processes.

The structure of chemical thermodynamics is based on the first two laws of thermodynamics. Starting from the first and second laws of thermodynamics, four equations called the "fundamental equations of Gibbs" can be derived. From these four, a multitude of equations, relating the thermodynamic properties of the thermodynamic system can be derived using relatively simple mathematics. This outlines the mathematical framework of chemical thermodynamics.

HISTORY

In 1865, the German physicist Rudolf Clausius, in his *Mechanical Theory of Heat*, suggested that the principles of thermochemistry, e.g. the heat evolved in combustion reactions, could be applied to the principles of thermodynamics. Building on the work of Clausius, between the years 1873-76 the American mathematical physicist Willard Gibbs published a series of three papers, the most famous one being the paper *On the Equilibrium of Heterogeneous Substances*. In these papers, Gibbs showed how the first two laws of thermodynamics could be measured graphically and mathematically to determine both the thermodynamic equilibrium of chemical reactions as well as their tendencies to occur or proceed. Gibbs' collection of papers provided the first unified body of thermodynamic theorems from the principles developed by others, such as Clausius and Sadi Carnot.

J. Willard Gibbs - founder of *chemical thermodynamics*

During the early 20th century, two major publications successfully applied the principles developed by Gibbs to chemical processes, and thus established the foundation of the science of chemical thermodynamics. The first was the 1923 textbook *Thermodynamics and the Free Energy of Chemical Substances* by Gilbert N. Lewis and Merle Randall. This book was responsible for supplanting the chemical affinity with the term free energy in the English-speaking world. The second was the 1933 book *Modern Thermodynamics by the methods of Willard Gibbs* written by E. A. Guggenheim. In this manner, Lewis, Randall, and Guggenheim are considered as the founders of modern chemical thermodynamics because of the major contribution of these two books in unifying the application of thermodynamics to chemistry.

OVERVIEW

The primary objective of chemical thermodynamics is the establishment of a criterion for the determination of the feasibility or spontaneity of a given transformation. In this manner, chemical thermodynamics is typically used to predict the energy exchanges that occur in the following processes:

- Chemical reactions
- Phase changes
- The formation of solutions

The following state functions are of primary concern in chemical thermodynamics:

- Internal energy (U)
- Enthalpy (H)
- Entropy (S)
- Gibbs free energy (G)

Most identities in chemical thermodynamics arise from application of the first and second laws of thermodynamics, particularly the law of conservation of energy, to these state functions.

The 3 laws of thermodynamics:

- The energy of the universe is constant.
- In any spontaneous process, there is always an increase in entropy of the universe
- The entropy of a perfect crystal(well ordered) at 0 Kelvin is zero

CHEMICAL ENERGY

Chemical energy is the potential of a chemical substance to undergo a transformation through a chemical reaction or to transform other chemical substances. Breaking or making of chemical bonds involves energy or heat, which may be either absorbed or evolved from a chemical system.

Energy that can be released (or absorbed) because of a reaction between a set of chemical substances is equal to the difference between the energy content of the products and the reactants. This change in energy is called the change in internal energy of a chemical reaction. Where $\Delta U^\circ_{f\,\text{reactants}}$ is the internal energy of formation of the reactant molecules that can be calculated from the bond energies of the various chemical bonds of the molecules under consideration and $\Delta U^\circ_{f\,\text{reactants}}$ is the internal energy of formation of the product molecules. The change in internal energy is a process which is equal to the heat change if it is measured under conditions of constant volume (at STP condition), as in a closed rigid container such as a bomb calorimeter. However, under conditions of constant pressure, as in reactions in vessels open to the atmosphere, the measured heat change is not always equal to the internal energy change, because pressure-volume work also releases or absorbs energy. (The heat change at constant pressure is called the enthalpy change; in this case the enthalpy of formation).

Another useful term is the heat of combustion, which is the energy released due to a combustion reaction and often applied in the study of fuels. Food is similar to hydrocarbon fuel and carbohydrate fuels, and when it is oxidized, its caloric content is similar (though not assessed in the same way as a hydrocarbon fuel — see food energy).

In chemical thermodynamics the term used for the chemical potential energy is chemical potential, and for chemical transformation an equation most often used is the Gibbs-Duhem equation.

Chemical Reactions

In most cases of interest in chemical thermodynamics there are internal degrees of freedom and processes, such as chemical reactions and phase transitions, which always create entropy unless they are at equilibrium, or are maintained at a "running equilibrium" through "quasi-static" changes by being coupled to constraining devices, such as pistons or electrodes, to deliver and receive external work. Even for homogeneous "bulk" materials, the free energy functions depend on the composition, as do all the extensive thermodynamic potentials, including the internal energy. If the quantities $\{ N_i \}$, the number of chemical species, are omitted from the formulae, it is impossible to describe compositional changes.

Gibbs function or Gibbs Energy

For a "bulk" (unstructured) system they are the last remaining extensive variables. For an unstructured, homogeneous "bulk" system, there are still various *extensive* compositional variables $\{ N_i \}$ that G depends on, which specify the composition, the amounts of each chemical substance, expressed as the numbers of molecules present or (dividing by Avogadro's number = 6.023×10^{23}), the numbers of moles

$$G = G(T, P, \{N_i\}) .$$

For the case where only PV work is possible

$$dG = -SdT + VdP + \sum_i \mu_i dN_i$$

in which μ_i is the chemical potential for the i-th component in the system

$$\mu_i = \left(\frac{\partial G}{\partial N_i} \right)_{T,P,N_{j\neq i},etc.} .$$

The expression for dG is especially useful at constant T and P, conditions which are easy to achieve experimentally and which approximates the condition in living creatures

$$(dG)_{T,P} = \sum_i \mu_i dN_i .$$

Chemical Affinity

While this formulation is mathematically defensible, it is not particularly transparent since one does not simply add or remove molecules from a system. There is always a *process* involved in changing the composition; e.g., a chemical reaction (or many), or movement of molecules from one phase (liquid) to another (gas or solid). We should find a notation which does not seem to imply that the amounts of the components (N_i) can be changed independently. All real processes obey conservation of mass, and in addition, conservation of the numbers of atoms of each kind. Whatever molecules are transferred to or from should be considered part of the "system".

Consequently, we introduce an explicit variable to represent the degree of advancement of a process, a progress variable ξ for the *extent of reaction* (Prigogine & Defay, p. 18; Prigogine, pp. 4–7; Guggenheim, p. 37.62), and to the use of the partial derivative $\partial G/\partial\xi$ (in place of the widely used "ΔG", since the quantity at issue is not a finite change). The result is an understandable expression for the dependence of dG on chemical reactions (or other processes). If there is just one reaction

$$(dG)_{T,P} = \left(\frac{\partial G}{\partial \xi}\right)_{T,P} d\xi.$$

If we introduce the *stoichiometric coefficient* for the *i-th* component in the reaction

$$\nu_i = \partial N_i / \partial \xi$$

which tells how many molecules of *i* are produced or consumed, we obtain an algebraic expression for the partial derivative

$$\left(\frac{\partial G}{\partial \xi}\right)_{T,P} = \sum_i \mu_i \nu_i = -\mathbb{A}$$

where, (De Donder; Progoine & Defay, p. 69; Guggenheim, pp. 37,240), we introduce a concise and historical name for this quantity, the "affinity", symbolized by **A**, as introduced by Théophile de Donder in 1923. The minus sign comes from the fact the affinity was defined to represent the rule that spontaneous changes will ensue only when the change in the Gibbs free energy of the process is negative, meaning that the chemical species have a positive affinity for each other. The differential for G takes on a simple form which displays its dependence on compositional change

$$(dG)_{T,P} = -\mathbb{A}\, d\xi\,.$$

If there are a number of chemical reactions going on simultaneously, as is usually the case

$$(dG)_{T,P} = -\sum_k \mathbb{A}_k\, d\xi_k\,.$$

a set of reaction coordinates { ξ_j }, avoiding the notion that the amounts of the components (N_i) can be changed independently. The expressions above are equal to zero at thermodynamic equilibrium, while in the general case for real systems, they are negative because all chemical reactions proceeding at a finite rate produce entropy. This can be made even more explicit by introducing the reaction *rates* $d\xi_j/dt$. For each and every *physically independent process* (Prigogine & Defay, p. 38; Prigogine, p. 24)

$$\mathbb{A}\dot{\xi} \leq 0.$$

This is a remarkable result since the chemical potentials are intensive system variables, depending only on the local molecular milieu. They cannot "know" whether the temperature and pressure (or any other system variables) are going to be held constant over time. It is a purely local criterion and must hold regardless of any such constraints. Of course, it could have been obtained by taking partial derivatives of any of the other fundamental state functions, but nonetheless is a general criterion for ($-T$ times) the entropy production from that spontaneous process; or at least any part of it that is not captured as external work. (See *Constraints* below.)

We now relax the requirement of a homogeneous "bulk" system by letting the chemical potentials and the affinity apply to any locality in which a chemical reaction (or any other process) is occurring. By accounting for the entropy production due to irreversible processes, the inequality for dG is now replaced by an equality

$$dG = -SdT + VdP - \sum_k \mathbb{A}_k \, d\xi_k + W'$$

or

$$dG_{T,P} = -\sum_k \mathbb{A}_k \, d\xi_k + W'.$$

Any decrease in the Gibbs function of a system is the upper limit for any isothermal, isobaric work that can be captured in the surroundings, or it may simply be dissipated, appearing as T times a corresponding increase in the entropy of the system and/or its surrounding. Or it may go partly toward doing external work and partly toward creating entropy. The important point is that the *extent of reaction* for a chemical reaction may be coupled to the displacement of some external mechanical or electrical quantity in such a way that one can advance only if the other one also does. The coupling may occasionally be *rigid*, but it is often flexible and variable.

Solutions

In solution chemistry and biochemistry, the Gibbs free energy decrease ($\partial G/\partial \xi$, in molar units, denoted cryptically by ΔG) is commonly used as a surrogate for ($-T$ times) the entropy produced by spontaneous chemical reactions in situations where there is no work being done; or at least no "useful" work; i.e., other than perhaps some $\pm\, PdV$. The assertion that all *spontaneous reactions have a negative* G is merely a restatement of the fundamental thermodynamic relation, giving it the physical dimensions of energy and somewhat obscuring its significance in terms of entropy. When there is no useful work being done, it would be less misleading to use the Legendre transforms of the entropy appropriate for constant *T*, or for constant *T* and *P*, the Massieu functions $-F/T$ and $-G/T$ respectively.

NON-EQUILIBRIUM SYSTEM CONSTRAINTS

Generally the systems treated with the conventional chemical thermodynamics are either at equilibrium or near equilibrium. Ilya Prigogine developed the thermodynamic treatment of open systems that are far from equilibrium. In doing so he has discovered phenomena and structures of completely new and completely unexpected types. His generalized, nonlinear and irreversible thermodynamics has found surprising applications in a wide variety of fields.

The non equilibrium thermodynamics has been applied for explaining how ordered structures e.g. the biological systems, can develop from disorder. Even if Onsager's relations are utilized, the classical principles of equilibrium in thermodynamics still show that linear systems close to equilibrium always develop into states of disorder which are stable to perturbations and cannot explain the occurrence of ordered structures.

Prigogine called these systems dissipative systems, because they are formed and maintained by the dissipative processes which take place because of the exchange of energy between the system and its environment and because they disappear if that exchange ceases. They may be said to live in symbiosis with their environment.

The method which Prigogine used to study the stability of the dissipative structures to perturbations is of very great general interest. It makes it possible to study the most varied problems, such as city traffic problems, the stability of insect communities, the development of ordered biological structures and the growth of cancer cells to mention but a few examples.

System Constraints

In this regard, it is crucial to understand the role of walls and other *constraints*, and the distinction between *independent* processes and *coupling*. Contrary to the clear implications of many reference sources, the previous analysis is not restricted to homogeneous, isotropic bulk systems which can deliver only PdV work to the outside world, but applies even to the most structured systems. There are complex systems with many chemical "reactions" going on at the same time, some of which are really only parts of the same, overall process. An *independent* process is one that *could* proceed even if all others were unaccountably stopped in their tracks. Understanding this is perhaps a "thought experiment" in chemical kinetics, but actual examples exist.

A gas reaction which results in an increase in the number of molecules will lead to an increase in volume at constant external pressure. If it occurs inside a cylinder closed with a piston, the equilibrated reaction can proceed only by doing work against an external force on the piston. The extent variable for the reaction can increase only if the piston moves, and conversely, if the piston is pushed inward, the reaction is driven backwards.

Similarly, a redox reaction might occur in an electrochemical cell with the passage of current in wires connecting the electrodes. The half-cell reactions at the electrodes are constrained if no current is allowed to flow. The current might be dissipated as joule heating, or it might in turn run an electrical device like a motor doing mechanical work. An automobile lead-acid battery can be recharged, driving the chemical reaction backwards. In this case as well, the reaction is not an independent process. Some, perhaps most, of the Gibbs free energy of reaction may be delivered as external work.

The hydrolysis of ATP to ADP and phosphate can drive the force times distance work delivered by living muscles, and synthesis of ATP is in turn driven by a redox chain in mitochondria and chloroplasts, which involves the transport of ions across the membranes of these cellular organelles. The coupling of processes here, and in the previous examples, is often not complete. Gas can leak slowly past a piston, just as it can slowly leak out of a rubber balloon. Some reaction may occur in a battery even if no external current is flowing. There is usually a coupling coefficient, which may depend on relative rates, which determines what percentage of the driving free energy is turned into external work, or captured as "chemical work"; a misnomer for the free energy of another chemical process.

REFERENCES

1. Ott, Bevan J.; Boerio-Goates, Juliana (2000). Chemical Thermodynamics – Principles and Applications. Academic Press. ISBN 0-12-530990-2.
2. Clausius, R. (1865). The Mechanical Theory of Heat – with its Applications to the Steam Engine and to Physical Properties of Bodies. London: John van Voorst, 1 Paternoster Row. MDCCCLXVII.
3. Klotz, I. (1950). Chemical Thermodynamics. New York: Prentice-Hall, Inc.

Index